Reviews and critical articles covering the entire field of normal anatomy (cytology, histology, cyto- and histochemistry, electron microscopy, macroscopy, experimental morphology and embryology and comparative anatomy) are published in Advancesin Anatomy, Embryology and Cell Biology. Papers dealing with anthropology and clinical morphology that aim to encourage cooperation between anatomy andrelated disciplines will also be accepted. Papers are normally commissioned. Original papers and communications may be submitted and will be considered for publication provided they meet the requirements of a review article and thus fit into the scope of "Advances". English language is preferred.
It is a fundamental condition that submitted manuscripts have not been and will not simultaneously be submitted or published elsewhere. With the acceptance of a manuscript for publication, the publisher acquires full and exclusive copyright for all languages and countries.
Twenty-five copies of each paper are supplied free of charge.

Manuscripts should be addressed to

209
Advances in Anatomy, Embryology and Cell Biology

For further volumes:
http://www.Springer.com/series/102

Katharina Spanel-Borowski

Footmarks of Innate Immunity in the Ovary and Cytokeratin-Positive Cells as Potential Dendritic Cells

With 33 figures

Katharina Spanel-Borowski
University of Leipzig
Institute of Anatomy
Liebigstr. 13
04103 Leipzig
Germany
Katharina.Spanel-Borowski@medizin.uni-leipzig.de

ISSN 0301-5556
ISBN 978-3-642-16076-9 e-ISBN 978-3-642-16077-6
DOI 10.1007/978-3-642-16077-6
Springer Heidelberg Dordrecht London New York

Cover design: WMXDesign GmbH, Heidelberg, Germany

Printed on acid-free paper

Springer is part of Springer Science+Business Media (www.springer.com)

Preface

The cyclic ovary can be seen as a site of tissue damage, repair and precisely controlled tissue homeostasis, as long ovulation and luteolysis can be compared with acute and chronic inflammation. Innate immunity appears to be a powerful force in the endocrine system, representing a novel concept. In this monograph, remarkable evidence is given for the immune-privileged ovary being an implant on the chicken chorioallantoic membrane with a non reactive mesenchyme. Mild to severe tissue damages due to follicular atresia, follicular rupture, or intraovarian oocyte release do no lasting harm. The most exciting part relates to the analysis of cytokeratin-positive (CK^+) cells, comparing the fate mapping of this cell type from the fetal ovary to the adult organ. Findings on toll-like receptor 4 regulation and interferon-γ-dependent positive effects indicate that CK^+ cells from human preovulatory follicles and bovine corpora lutea have similarities with nonlymphoid dendritic cells, a discovery that has the makings of a top story in basic and clinical research on the ovary.

Katharina Spanel-Borowski

Abstract

This monograph introduces innate immunity as a second force in the ovary in addition to the endocrine system. Innate immunity appears to orchestrate follicular atresia, follicle rupture, and follicle transformation into a corpus luteum (CL) and CL regression through sterile inflammation and tissue repair. The concept is new. It centers on cytokeratin-positive (CK^+) cells being recognized as a potential non lymphoid dendritic cell (DC) type. Part I describes morphological aspects of immune privilege starting with hamster ovary implants into the chicken chorioallantoic membrane with a non reactive mesenchyme. Follicular atresia and follicle rupture correspond to mild and moderate tissue damage in ovaries of small rodents and rabbits. Superovulations cause severe tissue damage through intra-ovarian oocyte release with follicle wall remnants in oedema, rupture of vessel walls and thrombosis. The complement system and neuropeptides might play regulatory roles. Part IIa analyzes intact ovaries (cows, human) for the appearance of CK^+ cells. In the fetal ovary, sex cords give rise to CK^+ cells in primordial follicles. In the adult ovary, CK^+ cells are absent in preantral follicles and reappear in mature and regressing follicles. In the CL of early development, steroidogenic CK^+ cells build a peripheral zone in the previous granulosa cell layer, and are uniformly distributed in the following stages. A microvascular CK^+ cell type is seldom found. Part IIb characterizes the morphology and function of CK^+ cells in vitro. Isolated from human preovulatory follicles, the epithelioid CK^+ granulosa cell subtype regulates TLR4 and CD14 at 36 h of treatment with oxidized lipoprotein (oxLDL, 150 μg/ml); non-apoptotic cell death and the increase of reactive oxygen species occur. In contrast, the CK-negative (CK^-) granulosa cell type regulates the lectin-like oxLDL receptor 1 (LOX-1) and survival autophagy under oxLDL stimulation. Isolated from bovine CL, the epithelioid CK^+ cell type 1 is disclosed as a microvascular cell type with a single non-motile cilium. The microvascular CK^+ type strongly upregulates intercellular contacts under treatment with interferon-γ (IFN-γ). In the CK^- cell type 5 of granulosa cell-like appearance, IFN-γ treatment supports cell proliferation, N-cadherin upregulation, and the dramatic increase in major histocompatibility complex II peptides (MHC II) by 80-fold compared to basal levels. Type 5 could have been converted from the steroidogenic CK^+ cell type. We summarize and conclude: CK^+ granulosa cells express functionally active TLR4, which sense danger signals, such as oxidative stress in preovulatory follicles, and trigger inflammatory and immunoregulatory pathways. The final outcome regulates follicle rupture and transformation into CL. Luteolysis could start by danger-sensing through the microvascular CK^{+} type 1 cells and the DC-like type 5 cells, both sensitive to IFN-γ. The future will witness a novel strategy in the therapy of ovarian disorders such as anovulations, luteal phase insufficiency and autoimmune failures.

Acknowledgements

This work relies on the support by many persons being good company at different German and Swiss locations. My sincere thanks go to Sonja Braumüller (Ulm), Ursula Almert, Waltraut Maaß (Lübeck), Markus Saxer (Basel), Annemarie Brachmann, Judith Craatz, Constanze Franke, Angela Ehrlich, Silke Kiessling, and in particular to Claudia Merkwitz (all from Leipzig) for technical skills and devotion to the experiments. My former graduate students from Ulm and Lübeck contributed important findings to cell proliferation, to intraovarian oocyte release, and to the endothelial cell subtype matrix (Drs. Ch. Heiss, Gerhild Rix, Andrea Löseke, Sabine Thor-Wiedemann, Prof. U. Böcker), to NCAM presence in luteal cell types (Prof. A. Mayerhofer) and to ectocytosis of CK^+ cells as an unusual form of secretion (Drs. Gudrun Herrmann, Hannah Missfelder-Lobos). The vascular corrosion casts of hamster ovaries is attributed to the good collaboration with Prof. W. Amselgruber at the start of our professional life. My next door neighbor at Lübeck, Dr. K.-W. Wolff, knew the staining technique for the single cilium of the CK^+ cells. Prof. K. Ley came as scientific assistant from Berlin to Lübeck to conduct leucocyte adhesion with the five phenotypes. Leucocyte subtype localization and freeze fracture studies in CK^+ cells were conducted by Drs. Patricia and Ch. Rahner at Basel. The support of all of them is here deeply appreciated and their career in the western world admired. Graduate students (now Drs. Judith Bausenwein, Ulrike Heider, Katja Hummitzsch, S. Karger, F. Rohm, Cornelia Simchen, Daniela Koch, Winnie Kunzelmann, Eleni Valerkou, Constanze Vilser) and collaborators from the Institute of Anatomy at Leipzig (Drs. M. Bauer, Elke Brylla, Nicole Duerrschmidt, E. Lobos, Sabine Löffler, Ina Reibiger, Kristina Sass, Gerelsul Tscheuschilsuren, N. Tsikoliasz, Prof. Gabriela Aust) paved the way by analyzing immune cells and tachykinin expression in ruptured follicles as well as adhesion molecules, angiogenic factors, antioxidants and intercellular contacts in isolated cell subtypes from follicles, corpora lutea and the surface epithelium. Dr. Heike Serke was the first to show the oxLDL-dependent TLR4 regulation in CK^+ granulosa cells and the non-apoptotic cell death. The creative collaboration on CK^+ cells from the corpus luteum and the surface epithelium by Dr. A. Ricken is always remembered by shared times at Basel and Leipzig. The success of the recent work depended greatly on the regular and generous supply of follicle harvests from the Clinic of Reproductive Medicine at Leipzig (Verona Blumenauer, Dr. F.A. Hmeidan). Professor A. Krishna, Varanasi,

India, who was funded by the Alexander von Humboldt-Foundation, reinforced the group with his findings on albumin-positive granulosa cell subtypes and on neuropeptide expression in polycystic ovaries at Lübeck and at Leipzig. Prof. Rina Meidan, Rehovot, Israel, trusted the five different phenotypes from the bovine corpus luteum. The fruitful cooperation with Prof. J. Borlak, Frauenhofer Institute at Hannover, Germany, on the oxLDL-LOX-1 system in granulosa cells signified a big step forward in developing my concept. My teachers at Ulm, Lübeck, Basel and at San Antonio, Texas (Prof. Ch. Pilgrim, Prof. W. Kühnel, Prof. D. Sasse, Prof. R. Reiter) supported my own scientific development. My editor, Prof. Schiebler, friendly and patiently encouraged me to write the story on the ovary. Albert Rast, Leipzig, never lost patience in teaching me the digital world and looked after the perfect lay-out of the plates. Dr. Reinhard Spanel, my husband and critical listener to changing interpretations, provided clarity with his drawings and was always serious about my work. Last but not least, thanks go to the German Research Foundation for continuous financial support. The Swiss National Foundation and the Interdisciplinary Centre of Clinical Research at Leipzig funded parts of the work. In retrospect, it was a laborious, yet great journey with many roundabouts, finally leading to a novel field of ovarian research and to the unexpected comprehensive view of many of my own findings. I thank Leipzig for the deep experience.

Contents

Abbreviations

3β-HSD	3β-Hydroxysteroid dehydrogenase
AP1	Activator protein-1
b-FGF	Basic fibroblast growth factor
C1q	Complement subunit 1q
C3	Complement factor 3
C3b	Complement subunit 3b
CAM	Chicken chorioallantoic membrane
CD	Cluster of differentiation
CK	Cytokeratin
CK^{+}/CK^{-}	Cytokeratin-positive/cytokeratin-negative
CL	Corpus luteum
ConA	Concanavalin A
DAG	Diacylglycerol
DAMP	Damage-associated molecular pattern
DAPI	4′,6-Diamidino-2-phenylindole
DC	Dendritic cells
Dil-acLDL	Dil-acetylated low-density lipoprotein
DMEM	Dulbecco's modified Eagle medium
FasL	Fas ligand
Fc	Fragment crystallizable
FSH	Follicle-stimulating hormone
FVIIIr	Factor VIII-related antigen
g	Gravidation
GM-CSF	Granulocyte monocyte colony-stimulating factor
hCG	Human chorionic gonadotrophin
HE	Haematoxylin and eosin
HEPES	4-(2-Hydroxyethyl)-1-piperazineethanesulfonic acid
HMGB1	High mobility group box-1
HOPA	Haematoxylin-orange-g-phosphomolybdenic acid-aniline blue
HRP	Horseradish peroxidase
i.v.	Intravenous
ICAM	Intercellular adhesion molecule
IFN I	Interferon of type I (α and β)

IFN II	Interferon of type II (γ)
IgG, IGE	Immunoglobulin class G and E
IL-1, IL-6, Il-12	Interleukin-1, -6, -12
INIM	Innate immunity
IOR	Intra-ovarian oocyte release
IP3	Inositol 1,4,5-triphosphate
IRAKs	Interleukin-receptor-associated kinases
IRF3	Interferon-regulatory factor 3
IU	International unit
IVF	In vitro fertilization
JAK-STAT	Janus kinase-signal transducer and activator of transcription 3
KIT	Tyrosine kinase KIT receptor
LC3	Microtubule light chain 3
LH	Luteinizing hormone
LOX-1	Lectin-like oxidized low-density receptor 1
LPS	Lipopolysaccharide
MAPK/ERK	Mitogen-activated protein kinase/extracellular-signal-regulated kinase
MBL	Mannan binding lectin
MD-2	Molecule for LPS responsiveness of TLR4
MHC	Major histocompatiblity complex
MMP	Matrix metalloproteinases
mRNA	Messenger ribonucleic acid
Myd88	Myeloid differentiation factor 88, TIR domain-containing adaptor molecule
NCAM	Neuronal adhesion molecule
NF-κB	Nuclear factor-κB
NGF	Neurotrophic growth factor
NK	Natural killer cell
NK-1R, -2R, -3R	Neurokinin-1/2/3 receptor
NKA, NKB	Neurokinin A and B
nLDL	Normal low-density lipoprotein
oxLDL	Oxidized low-density lipoprotein receptor
P450scc	Cholesterol side chain cleavage enzyme
PAMP	Pathogen-associated molecular pattern
PCOS	Polycystic ovary syndrome
PGF_{2a} and PGE_2	Prostaglandin F_{2a} and E_2
PMSG	Pregnant mare serum gonadotrophin
PPT-I, PPT-II	Pre-protachykinin A and B
PRR	Pathogen-associated pattern recognition receptor
PTX	Pentraxin
RANTES	Regulated on activation, normal T cell expressed and secreted
ROS	Reactive oxygen species

rpm	Revolutions per minute
RT-PCR	Reverse transcription polymerase chain reaction
s.c.	Subcutaneous
SP	Substance P
StAR	Steroidgenic acute regulatory protein
TGF-β	Transforming growth factor β
TIR	Toll-IL-1 receptor domain
TLR	Toll-like receptors
TNF-α	Tumour-necrosis factor α
TRAIL	TNF-related apoptosis-inducing ligand
TRAM, SARM	TIR domain-containing adaptor molecules
TRIF	TIR-domain-containing adaptor protein inducing IFN-β
TRIF, TIRAP	TIR domain-containing adaptor molecules
VCAM-1	Vascular cell adhesion molecule-1
Wnt	Combination of Wg (wingless) and Int (oncogene int-1)
ZO-1	Tight junction protein 1/zona occludens 1
α-2M	α_2-Macroglobulin

Chapter 1
Background

The ovary is a dynamic organ, because of continuous processes of cell proliferation and regression. They are intense in the reproductive period characterized by the cyclic formation of the mature follicle, and its rupture as well as the development, maintenance and disappearance of the corpus luteum (CL; Oktem and Oktay 2008). The processes require coordination in cell dynamics to maintain tissue homeostasis. The balanced action between proliferation and cell death guarantees a constant ovarian size during the reproductive period. Proliferation and involution heavily depend on the changing influence of gonadotrophins and intra-ovarian regulators as effectors of the endocrine system (Craig et al. 2007; Devoto et al. 2009; Matsuda-Minehata et al. 2006). The immune system is also evolving as a powerful force. First indications on immunosurveillance of the ovulatory process have been recently reviewed (Richards et al. 2008; Liu et al. 2008). The concept is presently advanced by considering innate immunity (INIM) as the sovereign authority in control of ovarian processes. It is speculated that the ovary itself has "educated" INIM to respond to ovarian needs. In other words, the ovary appears to shape the resident INIM effectors. This INIM function by design belongs to the revolutionary reflections of self-recognition (Matzinger 2002, 2007). This author beautifully explains the danger model, where INIM is concerned about danger signals independent of whether they arise from self (dying cells) or from non-self (infectious invaders). The ovary is a unique physiological example of the danger model.

1.1
Innate Immunity, Toll-Like Receptors and Danger Signals

Innate immunity provides immediate protection against pathogens already in the newborn (Turvey and Broide 2010). Because no training of T cells occurs in the thymus to differentiate between self and non-self molecules, INIM seems to have no memory about previous contacts with pathogens. This shortcoming is excellently compensated in evolution by the generation of pathogen-associated pattern recognition receptors (PRRs). The transmembraneous receptors classically recognize clusters of lipoproteins, lipopolysaccharides (LPS), peptidoglycans, nucleic acids and mRNA of bacteria and viruses, altogether termed pathogen-associated molecular patterns (PAMPs). The multiligand PRRs with multipurpose signaling comprise the nucleotide-binding oligomerization domain-like receptors, retinoic acid-inducible gene-like receptors, C-type lectin-like receptors and the toll-like receptors (TLRs) family (Takeuchi and Akira 2010). The best studied PRR has up to 12 subtypes in the mouse with different cluster recognition capacities (O'Neill and Bowie 2007; O'Neill 2008; Takeda and Akira 2005). The complex TLR-dependent signaling pathway leads to immediate defense by release of pro-inflammatory and immunoregulatory cytokines such as interleukins (IL-1, IL-6, Il-12), tumor-necrosis factor α (TNF-α), and interferon of types I and II (IFN α and β, IFN-γ) and of chemokines. Attracted leucocytes comprise macrophages, neutrophils and eosinophils, whereby the subtypes differ in dependence of diseases. Most important, resident dendritic cells (DCs) are a main source for the production of TLR and a key cell in INIM function (Banchereau and Steinman 1998; Mellman and Steinman 2001; Steinman and Banchereau 2007). Immature DCs of plasmacytoid or lymphoid precursor lineages immigrate from the bone marrow either to the epithelium or to the stroma of organs, where they handle antigen endocytosis through Fc receptors, and process the antigen with the help of the major histocompatibility complex (MHC). Only terminally differentiated, mature DCs express high MHC II peptides on the cell surface together with co-stimulatory molecules. Mature DCs are more effective as antigen-presenting cells than monocytes and macrophages, because DCs show MHC II peptides at significantly higher levels. Maturation of DCs is associated with hundreds of upregulated genes, just to name *IFN-I* (antiviral activity) and *Il-12* (differentiation of naïve T cells). Mature DCs activate natural killer cells (NK cells or large granular lymphocytes). Activated NK cells induce lysis of harmful cells through the killing activating receptor, provided that MHC I as killing inhibiting receptor is absent on the tackled cells. Mature DCs are considered to be mobile sentinels, because they transport the sampled antigen into regional lymph nodes. There, DCs stimulate B cells and drive naïve T cells into T-helper lineages, which in turn respond with the production of powerful cytokines, such as IL-1, TNF-α and IFN-γ, in a positive feed-back loop. The dialogue between mature DCs and lymphocytes points to the strong connexion of INIM with adaptive immunity (Iwasaki and Medzhitov 2010), which, in eons of evolution, appeared millions of years after INIM presence and function (Turvey and Broide 2010).

Innate immunity functions by barriers (Mayer 2009; Turvey and Broide 2010). First, the anatomical barrier protects against the outside, being the atmosphere for the skin or the mucus-rich environment for the mucosa lining hollow organ systems such as the broncho-pulmonary tract, the gastro-intestinal tract and the urogenital tract. The mucosa can provide cilia and fluid current for eliminating unhealthy invaders. The fluid also contains unspecific anti-pathogenic factors such as defensins, surfactant A and lysozyme for breakdown of the bacterial wall. Second, when the anatomical barrier cannot effectively cope with infectious invaders, the humoral barrier provides specific help with a wide repertoire of immunoregulatory cytokines (IFN I and II, IL-1, and TNF-α). The complement cascade is activated and potent serum factors are generated in response to the invaders. Some factors exert influence on the coagulation system. Vascular permeability increases, oedema form, thrombi are noted, and leucocyte subtypes are recruited for phagocytosis. All are well-known signs of acute inflammation. The recruited leucocytes belong to the third barrier responsible for the main line of defense by phagocytosis and killing of invaders. Phagocytic cells exert their influence through Fc receptors, which bind to the Fc region of IgG-coated bacteria, by receptors for the complement subunit 3b (C3b), and by scavenger receptors that recognize polyanionic surfaces of bacteria. Phagocytosis consumes glucose and oxygen leading to the respiratory burst with the production of reactive oxygen species (ROS) in a lysosome/myeloperoxidase-dependent or -independent manner. The third barrier also comprises mast cells and NK cells. Altogether, INIM reacts to unwanted signals on the very spot, and provides prodigious possibilities to modify the battlefield for leucocyte recruitment, angiogenesis and connective tissue changes. Barriers of INIM vary between healthy organs. Cilia of the respiratory epithelium reinforce the mucosal barrier in the broncho-pulmonary tract whereas, in the gastro-intestinal tract, dense leucocyte infiltrates in the lamina propria indicate the physiological answer against microbes. Inflammatory bowel diseases could depend on aberrant mucosal responses (van Lierop et al. 2009). Disorders of INIM also become relevant to the female genital tract as a cause of sexually transmitted diseases (Horne et al. 2008). In the ovary, INIM barriers seem to change in function for follicular atresia, follicle rupture, and formation and regression of the CL, because inflammatory cell patterns differ as explained below.

Non-specific immunity can no longer be the alternate term for INIM, because it confers great benefits to the body and supports healing of diseases (Turvey and Broide 2010). Novel and exciting concepts have been put forward that INIM is less concerned with recognition of PAMPs, but rather in the reception of danger/damage-associated molecular patterns (DAMPs) (Bianchi 2007; Matzinger 2002, 2007; Rock et al. 2010). The ultimate aim is to completely restore the integrity of architecture and function after organ damage. When danger signals come from outside through infectious invaders or by mismatched organ transplantation, the concept is still valid that the body discriminates exogenous non-self molecules. Yet, danger signals can be physiologically generated from inside by dying cells such as in the thymus and the gut mucosa or by tissue damage after surgical intervention. Cell and tissue damages cause the release of endogenous, non-foreign danger signals also called alarmins (Bianchi 2007; Rock et al. 2010). They comprise, e.g., Il-1α, the S100 calcium-binding family, acute phase proteins such as heat shock proteins and amyloid peptides, hyaluronan fragments, high mobility group box-1 (HMGB1), and uric acids. Alarmins are passively released from dying cells or actively secreted by stressed cells. Because many alarmins miss a signal peptide and the protein is thus leaderless, they are secreted by a non-classical exosomal pathway (Foell et al. 2007; Klune et al. 2008; Sandri et al. 2008; Zhan et al. 2009).

It lacks the canonical endoplasmic reticulum/Golgi protein trafficking, but requires active caspase-1 (Keller et al. 2008). Ischaemia and reperfusion of the liver leads to the release of HMGB1 from dying cells, which promotes the TLR4-dependent inflammation cascade through DCs (Tsung et al. 2007). Thermal skin injury significantly induces TLR-dependent production of IL-1β, IL-6 and TNF-α by spleen cells (Paterson et al. 2003). These cytokines also peak in the preovulatory follicle (Brännström et al. 1994a). In considering the presence of hyaluronan fragments in mature follicles (Shimada et al. 2008), alarmins could become highly relevant to the biology of the ovary.

1.2 The Ovary, Tissue Remodeling and Immune Privilege

In the past, the ovary has been a nice steady-state model to examine the dynamics of follicular growth and atresia by autoradiography after thymidine incorporation (Gougeon 1993; Zeleznik 1993). Yet no insights into the homeostasis of the interstitial cortex, which lacks proliferating and classical apoptotic cells (Spanel-Borowski et al. 1984), have been available until now. The old assumption that gonadotrophins and sex steroids exclusively control the status of hormone-receptive ovarian cells has been extended to intra-ovarian regulators (Adashi 1994; Adashi and Leung 1993). In the meantime, a wide knowledge exists about inflammatory cytokines and chemokines as well as their complex interactions during the life cycle of follicles and of CL (Hussein 2005; Craig et al. 2007; Devoto et al. 2009; Matsuda-Minehata et al. 2006). A comprehensive gene analysis has revealed novel immunoregulators in the preovulatory follicle (Espey 2006; Richards et al. 2002). Over the last decade, the mechanisms have been in focus as to how cell death occurs in regressing antral follicles and during luteolysis. It is not the lack of hormones and cytokines which lead to cell decay in the first line, but rather the activation of deleterious signaling cascades. One part of the cell disposal is decided by Fas–Fas ligand (FasL)-dependent apoptosis followed by immigration of macrophages for phagocytosis (Hussein 2005; Stocco et al. 2007; Wu et al. 2004). Another part of disposal apparently operates through necrosis and cell-death autophagy (Del Canto et al. 2007; Serke et al. 2009; Van Wenzel et al. 1999). Failed clearance of dead cells promotes inflammatory responses by stimulation of TLRs (Peng et al. 2007).

Tissue remodeling in the ovary is associated with temporal and spatial renewal of the microvascular bed. Angiogenesis is intense in dominant and preovulatory follicles as well as in the developing CL (Fraser and Wulff 2001, 2003; Reynolds et al. 2000). Areas with low oxygen levels obviously activate hypoxia-inducible factors that form heterodimers and lead to transcriptional regulation of vascular endothelial cell growth factor (VEGF) in granulosa cells and in the developing CL (Kim et al. 2009; Nishimura and Okuda 2010; Schams and Berisha 2004). The factor induces capillary sprouting from the thecal cell layer towards the avascular granulosa cell layer. The simultaneously produced angiopoietin-2 is required for the stabilization of the microvascular bed through Tie-2 receptors (Fraser and Duncan 2005; Peters et al. 2004; Stouffer et al. 2007). Other angiogenic/growth factors such as the basic fibroblast-growth factor (b-FGF) and insulin-like growth factor complete the maturation of the capillary network (Berisha and Schams 2005). In parallel, luteinization of granulosa and thecal cells is triggered by the pituitary-derived luteinizing hormone (LH) and further developed by local regulators such as oxytocin, endothelins, prostaglandins and progesterone itself (Berisha and Schams 2005; Devoto et al. 2009; Niswender et al. 2000; Stocco et al. 2007). Tissue remodeling also involves the timely precise sequence of upregulated selectins on endothelial cells as well as of integrin and chemokine receptors on leucocytes, determining adhesion and recruitment of leucocyte subtypes into follicles and CL, respectively (Brännström and Enskog 2002; Rohm et al. 2002). Leucocytes are selected in such a way that the population adapts to different demands. In preovulatory follicles and the forming CL, macrophages secrete metalloproteinases (MMPs) for the degradation of the extracellular matrix, and immigrated eosinophils might be helpful for angiogenesis (Munitz and Levi-Schaffer 2004; Wu et al. 2004). T cells represent a minority, whereas macrophages and segmented leucocytes are the majority (Brännström et al. 1994b; Best et al. 1996). Collectively, extracellular matrix degradation, angiogenesis and segmented leucocyte accumulation in the periovulatory body compares with a physiologal inflammation after wounding (Espey 1994, 2006; Medzhitov 2008, 2010a). The acute inflammatory response is induced by tissue stress during follicle rupture. Pro- and anti-inflammatory mediators regulate the event from damage to tissue repair within the current ovarian cycle. In contrast, removal of a CL requires several ovarian cycles, and the histological cell pattern corresponds with chronic inflammation involving macrophages for phagocytosis of dying endocrine and endothelial cells (Wu et al. 2004). Altogether, tissue remodelling in the ovary requires a complex inflammatory network, and the flexibility to respond to changing demands.

Under normal circumstances, an intact ovary is ready for the next cycle. This signifies that the ovary compensates well for repetitive tissue injuries caused by cyclic ovulations and the huge “garbage disposal” during luteolysis. Thus, the ovary should exert a strong and effective surveillance of tolerance. Indeed, the ovary survives after grafting and thus belongs to the immune-privileged organs such as the brain, the eye, the testis and the pregnant uterus (Ferguson et al. 2002; Krohn 1977). Immune-privileged organs are endowed with a special immunological status to restrict inflammation and deleterious consequences to exogenous danger signals. Major histocompatiblity complex Ib molecules are weakly expressed allowing the attack of cytotoxic T cells through the killing activating receptor that otherwise is inhibited by MHC Ib. Additionally, cell-membrane-bound molecules induce apoptosis of activated T cells, neutrophils and macrophages through the cytokine FasL and TRAIL (TNF-related apoptosis-inducing ligand). Complement regulatory proteins such as CD46 and CD56 inhibit the complement cascade (Niederkorn 2006). In the case of disorders, the function of organs is threatened, leading, e.g., to rejection of the fetus from the pregnant uterus. However, immunosurveillance appears to differ in outcome between the immune-privileged organs when considering their responses to danger signals. Brain and eye show devastating consequences to injury-induced inflammation, because the two organs have limited capacity for tissue regeneration after injury. Yet in the ovary, strong inflammatory changes in the follicle under rupture are followed by excellent tissue repairs. Thus, the INIM of the ovary must have been well educated according to a consideration that each organ could be in control of danger signals (Matzinger 2007).

1.3 Design

Follicular atresia should be a more minor threat for the ovary than ruptures of preovulatory follicles in cyclic and superovulated mammalian ovaries. Consequently, the morphological answers should differ with mild, moderate and severe tissue damage. For this reason, my own findings with rodent ovaries, which were studied many years ago, have been checked with the aim of detecting morphological footmarks of INIM function. Under the hypothesis that INIM is the coordinating force, a specialized cell population should be characterized like the Langerhans cells in the epidermis. Granulosa cells with cytokeratin (CK) expression, in the following called CK-positive (CK^+) cells, have come into focus as promising candidates. The thorough morphological analysis of CK^+ cells in the fetal and adult ovary compares with a morphological and functional fate mapping of CK^+ cells in bovine and human ovaries. The characterization of CK^+ cell cultures derived from preovulatory follicles and from CL reveals criteria comparable to DCs. Future efforts will prove the still incomplete concept that CK^+ cells represent non-lymphoid DCs as powerful tools of ovarian INIM.

Chapter 2
Materials and Methods

Since Materials and Methods were described in detail in our publications, this brief survey refers to the species, and those points are elaborated which have been neglected and might be helpful to the reader. The technical expertise grew over the years. The first part preferentially relies on classical morphological techniques to study functional morphology of intact ovaries. Serial sections of superovulated ovaries from small laboratory animals were screened for the number of intact and regressing follicles and CL. The interest in angiogenesis during transformation of a mature follicle into a CL was the reason to work with ovarian cell cultures. Therefore, the second part focuses on the characterization and function of steroidogenic and endothelial cells isolated from preovulatory follicles and CL. Because rats and golden hamsters have a 4-day ovarian cycle, which is physiologically remote from humans, a switch in the "materials" occurred. Granulosa cell cultures with CK^+ and CK^- negative (CK^-) were established from preovulatory follicle harvests obtained from patients under in vitro fertilization (IVF) therapy, and five different phenotypes were isolated from the bovine CL. Protocols for single and double antigen localization by indirect immunohistology with the avidin–biotin–peroxidase-complex technique were established for paraffin sections of the intact ovary. The technique was extended to immunofluorescence localization in cell cultures. Methods for ultrastructure and scanning electron microscopy of monolayers were developed as well as cell proliferation studies and techniques of mRNA and protein analysis, (RT-PCR analysis, western blotting and flow cytometric analysis).

2.1
Cyclic and Superovulated Ovaries from Rats, Hamsters, and Rabbits as well as Canine, Bovine and Human Ovaries

Golden hamsters (*Mesocricetus auratus*), strains C-lac and Bom, were laboratory-reared and weaned at the age of 21 days. Additional animals were purchased from a commercial breeder; Sprague-Dawley rats were 21 days old, Beagle dogs were at different estrous cycle stages, and rabbits of the New Zealand breed were fertile at 3 months of age. Animals arrived 1–3 weeks ahead of time to get acclimated to the new environment before the onset of the experiments. The animals were kept at the local animal facilities of the Medical Faculty, where the animal room was air-conditioned (21 ± 2°C), windowless, with illumination from 06:00 to 20:00 hours. Rats and hamsters were housed in transparent plastic cages, 3–4 animals per cage, and with free access to food and water, while dogs and rabbits were individually caged.

In immature rats and golden hamsters, superovulations were induced by subcutaneous (s.c.) injections of gonadotrophins to synchronize the ovarian cycle and to obtain time-related data (Löffler et al. 2004a; Spanel-Borowski and Heiss 1986; Spanel-Borowski et al. 1983b, 1986). The dose in international units (IU) of follicle-stimulating hormone (FSH) and LH per animal could be mild (10 IU follicle FSH + 10 IU LH), moderate (25 U PMSG + 25 IU LH), or high (50 IU PMSG + 25 IU LH), and it was correlated with low, moderate and high numbers of preovulatory and ruptured follicles. The injection schedule considered the circadian rhythm, starting at 08:00 hours with pregnant mare serum gonadotrophin (PMSG/FSH) and continuing 60 h later at 20:00 hours with LH. Ovulations, which first occurred 72 h later, i.e. on day 3 after PMSG, decreased in number until day 4 and even day 5 after PMSG (Fig. 2.1a). The CL disappeared on day 7, thus the regular 4-day-estrous cycle was extended by 3 days in superovulated hamster ovaries (Fig. 2.1b–e). For adult rabbits, which are reflex ovulators, the inducing dose was calculated per kg body weight: s.c. application of 35 IU pure FSH and intravenous (i.v.) 100 IU human chorionic gonadotrophin (hCG) 52 h later (Spanel-Borowski et al. 1986).

Bovine ovaries were dissected immediately after death of the cow at the local slaughterhouse, CL prepared, and functional stages assessed by macroscopical examination (Fig. 2.2a). The early developmental stage displayed the gyrated wall of a freshly ruptured follicle; it was 1 cm in diameter, and dark brown because of intense hyperaemia. The CL of the late developmental stage and of secretion was more than 2 cm in diameter, and consisted of soft tissue with colors from brown to weak-yellow; the distinctly yellow CL of early regression was becoming firm due to connective tissue increase. The CL of late regression was again small, firm and orange. The macroscopical diagnosis was confirmed by characteristic changes of the microvascular bed in ovaries fixed with 4% buffered formaldehyde, embedded in paraffin wax, sectioned and immunostained for the endothelial cell-specific factor VIII-related antigen (FVIIIr; Ricken et al. 1995). The capillary network was incomplete at the stage of development and was fully developed at the time of secretion. Capillaries decayed at the time of regression, whereas arterioles became abundant (Ricken et al. 1995; Bauer et al. 2001; Fig. 2.2b–d). To obtain fetal bovine ovaries, the uterus bicornis of pregnant cows was opened, and the fetus was removed. The gestational age was calculated by the crown–rump length (Tsikolia et al. 2009).

Human ovaries were obtained as paraffin-embedded samples from the archives of the Institute of Pathology, Medical Faculty, Leipzig. Fetal ovaries from abortions and miscarriages were allotted to the early, middle and late gestational periods. Cyclic ovaries with a developing CL were obtained from women between 19 and 38 years of age during operations for tubal pregnancy or cystadenoma (Löffler et al. 2000).

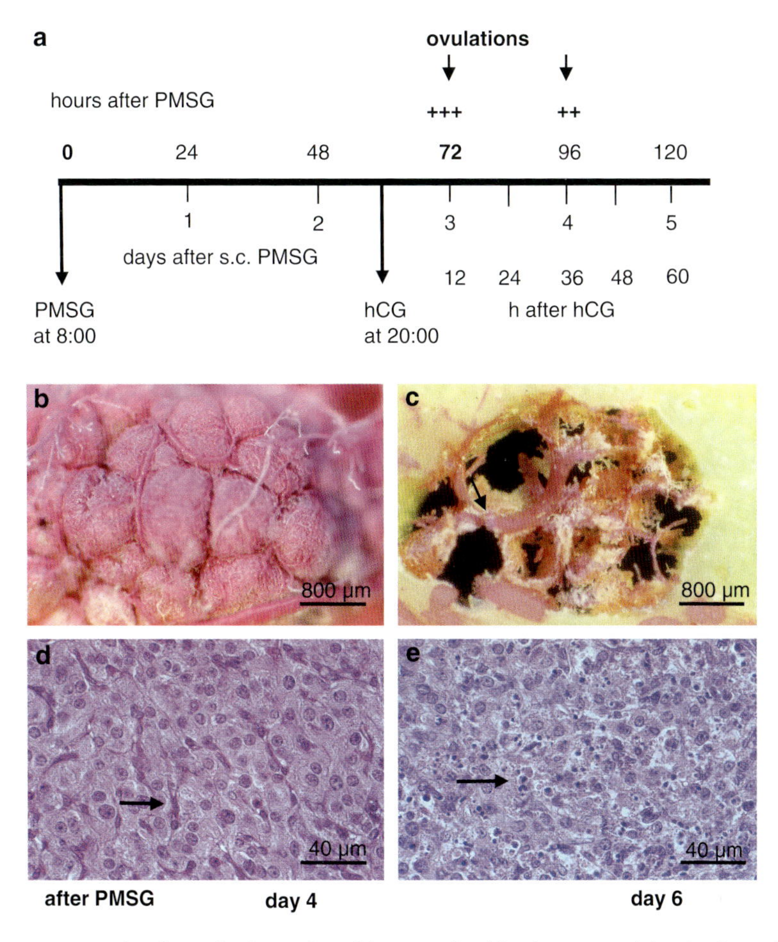

Fig. 2.1 Superovulated ovaries in 28-day-old rats and golden hamsters. (**a**) Injection scheme for superovulations. The s.c. dose per animal can be low (10 IU PMSG + 10 IU hCG), moderate (25 IU PMSG + 25 IU hCG), or high (50 IU PMSG + 25 IU hCG). (**b** and **d**) On day 4 after a high dose of PMSG, the vascular corrosion cast of hamster ovaries shows many CL, which consist of luteal and endothelial cells in the HE-stained sections (*arrow in* **d**). (**c** and **e**) On day 6, filling defects are noted in the cast accompanied by many apoptotic bodies in the histological section (*arrow in* **e**) and a decrease in the average ovarian weight from 150 mg to 30 mg (compare **b** and **c**). Adapted from Spanel-Borowski and Heiss (1986) and Spanel-Borowski et al. (1987)

2.1.1 Fibrinolytic Activity and Implants on the Chicken Chorioallantoic Membrane

The activity of plasminogen activator/fibrinolytic enzyme was localized in cryostat sections of superovulated ovaries (Spanel-Borowski 1987; Spanel-Borowski and

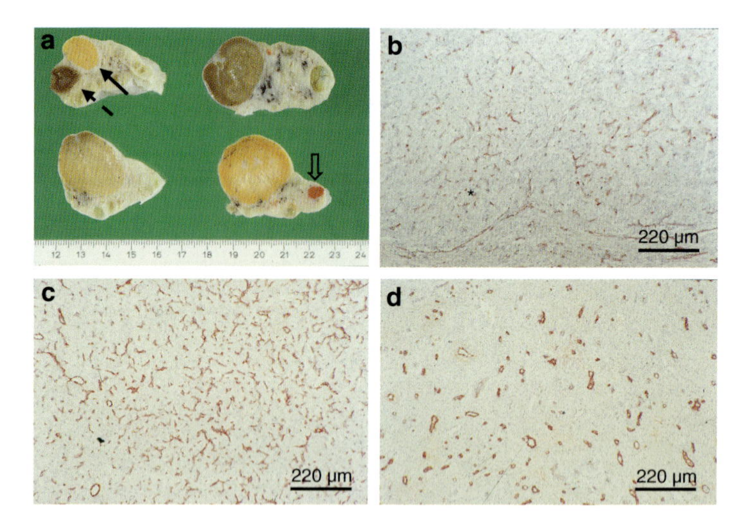

Fig. 2.2 Bovine CL in different stages of the estrous cycle either unfixed (**a**) or after fixation with 4% buffered formaldehyde for immunostaining of FVIIIr antigen to reveal the microvascular bed (**b**–**d**). (**a**) The CL in development is small and of brown color (*broken arrow*), whereas the CL in regression is *yellow* (*arrow upper left*). The CL of secretory stages are seen in the *upper right* and *lower left* corners. The CL of regression is large in the early stage and small in the advanced stage (*hollow arrow*). (**b**–**d**) The stage of development depicts more capillaries in the periphery of the CL (*asterisk*) than the centre (**b**). Capillaries are fully developed at the secretory stage (**c**). Capillaries disappear and thick-walled arterioles appear in the regressing stage (**d**). Adapted from Ricken et al. (1995) and Spanel-Borowski et al. (1997)

Heiss 1986). Two percent human fibrinogen solution (0.1 ml) rich in plasminogen was dropped onto a section and spread with the edge of another slide. The fibrinogen-slide was tilted to accumulate liquid in one corner, and 0.1 ml thrombin (25 IU) added followed by quick rotating movements. The fibrin film was stabilized in a moist chamber at 37°C for 10 min. Fibrinolytic areas were semi-quantitatively evaluated.

The implant experiments with cyclic and superovulated ovaries on the chorioallantoic membrane (CAM) were conducted with the intention of evaluating capillary sprouting from the ovary into the CAM (Spanel-Borowski 1989). We obtained fertilized chicken eggs at a local hatchery and kept them in an egg incubator at 37°C. On day 2 of incubation, 1 ml albumen was aspirated through a hole drilled in the blunt end; a 1.2-cm^2 window was cut in the shell of the egg's wide side, and closed with a cover slip with silicon grease. On day 9, cyclic ovaries and ovaries with mild and strong stimulation were gently applied until day 14. In comparison, the heart, the liver, the kidney, and the adrenal gland from immature and adult animals were used as implants. Implants were fixed in situ by Bouin's solution for 30 min.

2.1.2
Fixation, Staining and Counting Ovarian Structures

Ovaries were fixed in Bouin's solution or in 4% buffered formaldehyde overnight, embedded in paraffin wax, and sections of 7 μm serially cut at the longitudinal side. Routine stains used were haematoxylin and eosin (HE) as well as a tetrachrome (haematoxylin-orange-g-phosphomolybdenic acid-aniline blue, HOPA). We began to count proliferating cells in intact and regressing follicles in sections from dogs, which had been injected with (^{3}H)-thymidine; thymidine incorporation was revealed by autoradiography (Spanel-Borowski et al. 1984). The morphological observation on two forms of atresia and on the intra-ovarian oocyte release (IOR) led to extensive counting of follicular atresia, IOR, and of CL in completely cut cyclic and superovulated ovaries (Spanel-Borowski and Heiss 1986; Spanel-Borowski et al. 1982, 1983b). Sections of 7 μm thickness were collected into four alternate series at an interval of 28 μm (numbered 1a, 1b, 1c, 2a, 2b, 2c, etc.). The first series was stained with HE (numbered 1a, 2a, 3a, 4a, etc.). The second series was treated with the periodic-acid-Schiff reaction, and the rest were kept as reserve series. For rats and hamsters, every fourth out of a completely cut ovary was evaluated for antral/preovulatory follicles when the largest cross-section of the oocyte was seen. In considering the oocyte nucleus (approximately 28 μm in diameter) as "marker", redundancy in counting was avoided. The CLs were counted in every tenth section. The estrous cycle was staged by typical changes of the vaginal epithelium.

2.2
Cell Culture Subtypes from Follicles Derived from Patients Under IVF Therapy

For the granulosa cell subtypes, follicle harvests from women under IVF therapy were obtained from the Clinics of Reproductive Biology and Endocrinology, Dr. F. Hmeidan and colleagues, Leipzig, Germany. Patients had given written consent to the scientific project no. 116, which had been approved by the Committee for Ethics at the University of Leipzig. The aspirates from different follicles of one patient were pooled and centrifuged at 1,200 rpm for 3 min (Serke et al. 2010). Samples with clotted fluid were discarded. Harvests of red color were subjected to treatment with ammonium chloride solution to remove erythrocytes. The pellet was resuspended in 5 ml of 0.155 M NH_4Cl, 0.1 mM Na_2EDTA and 0.01 M $KHCO_3$ for 5 min. Then, 45 ml of phosphate-buffered saline was added before a 10-min centrifugation. Erythrocyte lysis was optional in yellow-colored harvests, because erythrocytes and leucocytes disappeared by efficient buffer washing the next day and every third day. A third way to remove the erythrocytes and the damaged cells was achieved with a discontinuous Percoll density gradient (Löffler et al. 2000). In a 15-ml conical plastic tube, 3 ml of 50% Percoll solution were covered with 7 ml cell harvest and centrifuged at 2,500 rpm for 7 min. The upper cell band was free of erythrocytes. The granulosa cells proliferated in Dulbecco's modified Eagle medium (DMEM) mixed 1 + 1 with Ham-F12 (Table 2.1). To the company-supplemented 15 mM 4-(2-hydroxyethyl)-1-piperazineethanesulfonic acid (HEPES), we added 22 mM $NaHCO_3$ and 5% fetal calf serum. Cell growth was improved by 10% Endothelial Cell Growth Medium MV (Promo Cell, C-22020; Heidelberg, Germany) and 10% serum. Commercial antibiotics were used. Within 5–8 days, granulosa cell cultures became confluent in 24-well-culture plates, and two subtypes were apparent, which were the epithelioid subtype with the expression of CK filaments and the fibroblast subtype. Because the subtypes seldom occurred as pure culture, further purification was done by mechanical colony transfer under the phase-contrast microscope (Spanel-Borowski and van der Bosch 1990; Fig. 2.3c). The serum-containing medium was removed with a HEPES buffer wash, and 200 μl of 0.05% trypsin solution applied for seconds until

Table 2.1 Simplified method for isolation of five phenotypes from the bovine CL

Reagents

Dulbecco's modified Eagle medium (DMEM) and nutrient mixture Ham-F12, powder mixture 1 + 1 (Invitrogen; cat. Nr. 42400-010) prepare fresh with sterile aqua dest., sterilize and keep at 4°C for 3 weeks because of decay in glutathione and Hepes

1 M $NaHCO_3$ stock solution, aliquot in 50 ml flask and store at 4°C until use

Fetal calf serum (FCS; Biochrom, Berlin, Germany)

Phosphate-buffered saline without calcium and magnesium, pH 7.1–7.2 (Sigma)

Preparation of medium with serum termed DF++

100 ml DMEM-F12

No supplementation with 15 mM Hepes because in the powder medium

+2.2 ml $NaCO_3$-stock for a 0.22 mM $NaHCO_3$

+5% FCS

Prepare fresh; concentrations must not be exact; pH 7.2 comes with buffer supplementation; no antibiotics are needed under sterile working conditions and thorough buffer washes

Collagen coating of 24-well culture plates

Stock solution (Pure cole; Inanmed Biomaterial, San Diego, Calif) at 4°C

Dilute 1:100, i.e. 1 ml add 100 ml DF+

Add 0.5 ml per well; incubate 2–24 h at 37°C

Remove solution immediately before use

Protocol

Select bovine ovaries for CL in specific functional stages

Rinse in 70% alcohol; dissect the CL from the capsule

Remove roughly 400 mg tissue from the peripheral region close to the previous ruptured surface epithelium

Transfer coarse pieces into 50-ml conical tubes, wash in 30 ml PBS at 1,200 rpm, 5 min

Mince pellet with a scalpel until a "paste" is obtained; wash with 20 ml DF+, two times at 1,200 rpm, 5 min for removal of dead cells; pipette vigorously to release cells from tiny fragments, use pipettes of decreasing size (20–2 ml), repeat 3× until supernatant becomes clear

Resuspend in DF+ and sieve through 150-μm and 72-μm pore size nylon meshes

After buffer wash, resuspend pellet in 12 ml DF++, add 0.5 ml per well

Incubate at 37°C and 5% CO_2 in air overnight; buffer wash, 2×, with DF+, 1×× with DF++

Observe colony development under the phase microscope the next day

Do colony transfer of a confluent island at 10 objective magnification

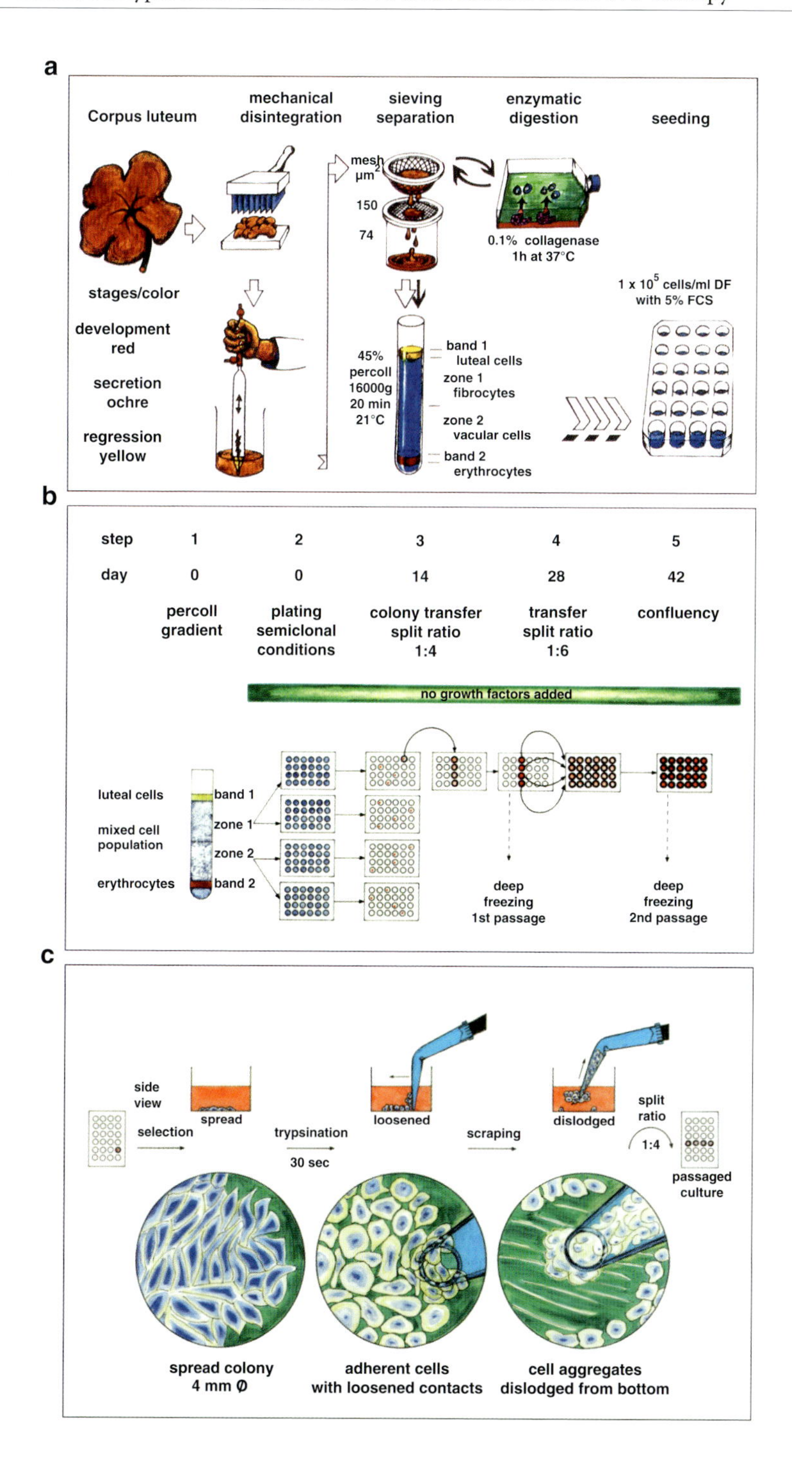
a
Corpus luteum
mechanical disintegration
sieving separation
enzymatic digestion
seeding
mesh µm²
150
74
0.1% collagenase 1h at 37°C
stages/color
development red
secretion ochre
regression yellow
1 x 10⁵ cells/ml DF with 5% FCS
45% percoll 16000g 20 min 21°C
band 1 luteal cells
zone 1 fibrocytes
zone 2 vacular cells
band 2 erythrocytes
b
step 1 2 3 4 5
day 0 0 14 28 42
percoll gradient
plating semiclonal conditions
colony transfer split ratio 1:4
transfer split ratio 1:6
confluency
no growth factors added
luteal cells
mixed cell population
erythrocytes
band 1
zone 1
zone 2
band 2
deep freezing 1st passage
deep freezing 2nd passage
c
side view
selection
spread
trypsination
30 sec
loosened
scraping
dislodged
split ratio
1:4
passaged culture
spread colony 4 mm Ø
adherent cells with loosened contacts
cell aggregates dislodged from bottom

intercellular contacts started to become loose. Leaving the culture with buffer alone for minutes also opened the intercellular contacts. The process was stopped with 1 ml of serum-containing medium, and the selected colony gently scraped off using the tip of a bent Eppendorf pipette. Cells were removed with serum-containing medium and immediately plated at a 1:4 split ratio in 24-well culture plates, with 1 ml serum-containing medium added for further enzyme inactivation. The purity of the cultures was confirmed by immunostaining for CK with the pancytokeratin antibody lu5 and by western blotting for CK8. For the oocyte-associated granulosa cells, cumulus oophorus complexes were incubated with hyaluronidase solution, and the oocyte with the corona radiata cells was removed before cultivating cumulus cells (Serke et al. 2009).

2.3 Cell Culture Subtypes from Bovine CL and Characterization

For subtypes derived from the CL, bovine ovaries were selected for CL stages in the laboratory (Ricken et al. 1995; Fig. 2.2a). The fresh preparation of the DMEM-Ham-F12 (Table 2.1) from powdered medium is recommended, because higher yields of colonies were noted than using the commercially available fluid medium. The supplementation with 5% fetal calf serum was sufficient. The whole CL was excised from the capsule and the tissue chopped with a scalpel and a cutting-device of 20 parallel razor blades (Fig. 2.3a). The tissue mince was vigorously pipetted for cell release, serially sieved through 150-μm and 72-μm^2 pore size nylon meshes, and washed. The final pellet was resuspended in 20 ml of isotonic 50% Percoll solution and, to obtain a continuous gradient, it was centrifuged at 16,000 rpm at room temperature for 20 min. The gradient developed an upper band with dominant luteal cells, a superior and inferior zone with fibrocytes and vascular cells, respectively, and with the erythrocytes at the bottom. The growing expertise showed no clear separation of subtypes, in particular in the upper zone, which was mixed with granulosa-like cells subsequently termed type 5 cells. Because the upper band contained many damaged cells, it was discarded. The fractions from the two zones were thoroughly washed to remove Percoll remnants before plating one fraction onto two plates of a 24-well culture plate (Fig. 2.3b). It had been coated with 1% collagen solution at 37°C, 2–24 h in advance. The yield of viable cells was low in the toluidine blue test so that fewer than 50 cells came into one well. The minority adhered and developed under semi-clonal conditions into colonies of five different phenotypes. Colonies with contact-inhibited growth

Fig. 2.3 The schematic procedure to isolate different phenotypes from the bovine CL. (**a**) Mechanical disintegration, sieving and further separation by a continuous Percoll gradient. (**b**) The time scale from cell plating under semiclonal conditions to deep-freezing of pure confluent cultures. (**c**) Colony transfer. Drawn by R. Spanel

allowed the mechanical selection by colony transfer (Fig. 2.3c). Contaminating fibroblasts were neglectable because of only mechanical CL disintegration. Pericytes were excluded by their non-contact inhibited growth. The problem of successful colony selection corresponded to the different proliferation rate of the five phenotypes whereby the subtype with CK expression grew moderately (Fenyves et al. 1994). Strict clonal conditions in a 98-well plate were unsuccessful in obtaining different phenotypes. Meanwhile, the method was simplified by discarding the Percoll gradient and by using a portion from the peripheral CL near the previous follicle ruptures area (Table 2.1). The portion of 400 mg tissue roughly replaced the yield from the whole CL. A mild neuronal tissue dissociation kit (Miltenyi Biotec, Bergisch Gladbach, Germany) did not improve the outcome. The exclusively mechanical digestion remained the method of choice.

Because the five phenotypes resembled endothelial cells and maintained a stable morphology in long-term culture, endothelial cell criteria were validated by light microscopy, such as monolayer appearance, spontaneous tubule formation, presence of FVIIIr antigen and uptake of Dil-acetylated low-density lipoprotein (Dil-acLDL; Spanel-Borowski and van der Bosch 1990). The cytoskeleton was immunostained for actin filaments, for CK, for intercellular contacts, and also for the extracellular fibrin matrix (Fenyves et al. 1993). The ultrastructure was investigated by transmission electron microscopy, scanning electron microscopy, and by freeze fracture for the intercellular junctions (Herrman et al. 1996; Ricken et al. 1996; Spanel-Borowski 1991).

Chapter 3
Footmarks of INIM

"See no evil, hear no evil and do no evil" is the appealing title of a review of immune-privileged organs (Niederkorn 2006). The statement perfectly fits the ovary, which faces regular damage from follicular atresia and cyclic follicle rupture as well as garbage disposal of the CL. The ovary hears and responds to danger signals by inflammatory processes (Espey 1994), yet it does no evil. By coordinated repair processes, the function of the ovary is perfectly restored. Intensities of responses appear to be tightly connected with the degree of danger signals, supporting the consideration that the tissue is in control (Matzinger 2007). Hence, immune management of follicular atresia is likely different compared to follicle rupture and CL regression. At the onset of the next ovarian cycle, all immune factors are reset for the next round. The concept that INIM is the major authority behind the endocrine throne is a novel approach to reveal the secret of how tissue homeostasis is handled. In the following, thoughts and findings by us and others are considered in the light of INIM functions in the ovary.

3.1 The Complement System as Danger Sensor in General and in the Ovary

The complement system with 30 serum and cell-surface proteins plays a role in INIM before evolution generates adaptive immunity (Endo et al. 2006). The powerful complement system comes in first by recognizing and transmitting threats into inflammatory responses (recruitment of leucocytes, vascular permeability, angiogenesis and clearance of damaged cells), all contributing to tissue integrity (Köhl 2006a, b). The complement system senses harmful structures and translates danger signals into organ-tailored INIM responses by danger sensors. The first complement subunit 1q (C1q) is a pattern recognition protein (danger sensor) which belongs to the TNF-ligand family and triggers the classical complement pathway (Endo et al. 2006; Ghai et al. 2007). The mannan binding lectin (MBL) and ficolin, both members of the collectin family with a collagen-like domain, represent danger sensors, which activate the lectin complement pathway through attached serine proteases. The C1q and MBL proteins transmit the danger signals into cellular responses by interaction with specific receptors on the cells, whereas C3, another danger sensor, is cleaved by convertases into subunits (C3a, C3b, and inhibitory C3b). Danger transmission depends on C3-subunit-specific receptors (Köhl 2006a, b). The activation cascade of C1q and C3 ultimately leads to humeral and cellular inflammatory responses, and to the membrane-attack pathway. C3a together with C4a and C5a belong to the powerful anaphylatoxins. The C5a receptor represents G-coupled cell membrane receptors with proinflammatory properties. The signaling spectrum comprises the mitogen-activated protein kinase/extracellular-signal regulated kinase (MAPK/ERK) cascade, phospholipase C, to generate the second messenger inositol 1,4,5-triphosphate (IP3) and diacylglycerol (DAG) for Ca^{2++} mobilization, as well the janus kinase (JAK)-signal transducer and activator of transcription 3 (STAT) pathway for cell proliferation. The C5a receptor signaling on connective tissue cells appears to exert a profibrotic role for tissue repair (Köhl 2006a, b).

Theoretically, the complement system might be shaped by the individual demands of the ovary to cope with tissue damage and, in parallel, to organize tissue repair. The importance of the complement system in ovarian function is being recognized. The *C1q* subunit mRNA in the cumulus cells of superovulated mice (Shimada et al. 2006) and the protein are found in cortical granules of fish oocytes (Mei et al. 2008), pointing to the evolutionarily old complement system. Ligands of the C1q subunit belong to the pentraxin (PTX)-family with members such as serum amyloid and C-reactive protein. The PTX are acute phase proteins, highly conserved in evolution, produced by, e.g., immune cells and endothelial cells in response to inflammatory signals and TLR contribution (Bottazzi et al. 2006). Mice gene-deficient in *ptx-3* are infertile, which is explained by a defective cumulus expansion. Infertility might also correspond with overactive inflammation in the preovulatory follicle, because experiments on intestinal ischaemia in *ptx-3* transgenic mice report an exaggerated inflammatory response (Souza et al. 2002). Although the liver is the main locus of complement production, follicle cells themselves might generate complement factors. An important contribution comes from a thorough analysis with modified proteomic techniques allowing the study of middle- and low-abundance proteins. In comparison with serum, the follicular fluid from IVF patients provides lower levels of complement 3 (C3) and C4a subunit and higher levels of C9 in addition to decreased haemolytic activities (Jarkovska et al. 2010). The findings signal reduced exchange of blood proteins through the follicle–blood barrier and the active involvement of the complement cascade in INIM function in the ovulatory event. The manifold increase of ficolin-3 precursors in the follicular fluid might point to the lectin pathway of complement activation (Matsushita 2009; Jarkovska et al. 2010). The increase of clusterin seems to act against complement/cytolytic activity in the preovulatory follicle, because clusterin inhibits the membrane attack complex C5b–C9. It is noteworthy that membrane-associated regulators of the complement system, such as membrane co-factor protein (CD46) and decay accelerating factor (CD55), which both inactivate C3 and CD4 convertases, are localized by immunostaining in human follicles and CL (Oglesby et al. 1996). The C4b-binding protein, another complement regulator, could play a role in connective tissue growth and resorption, because the β form of the protein co-localizes with fibroblast-like cells in the CL of advanced regression (Criado-Garcia et al. 1999). Altogether, the time has come to look closely at activating and inhibiting factors of the complement system in the ovary. The first step is to describe different degrees of tissue damage in the ovary as signs of danger responses.

3.2 Mild Danger with Mild Response

3.2.1 Implantation of the Ovary into the CAM

Before the discovery of angiogenic factors, the activity of capillary sprouting was assessed by implanting pieces of the bovine CL into the CAM and determining ingrowth of sprouts from the vascularized chorionic epithelium into the implant (Jakob et al. 1977). We also showed capillary sprouting in the absence of an inflammatory response in the CAM mesenchyme (Spanel-Borowski 1989). The whole ovary remained intact apart from a small necrotic zone close to the sprouting capillaries (Fig. 3.1a–d). In comparison, the CAM mesenchyme depicted a mild inflammatory response after implantation of the heart, whereas the liver lobe implant increased the reaction conspicuously (Fig. 3.1e–h), which was also noted for lung and kidney implants. Capillary sprouting was more severe in unstimulated ovaries from immature hamsters than from cyclic and superovulated ovaries. Thus, neither steroid hormones nor angiogenic factors could completely explain the acceptance of the ovarian implant. The CAM itself has a low histocompatibility barrier, because immunocompetence of the chicken embryo is developed at the time of hatching. It thus appears that the immune status of the ovary itself slows down the CAM responses. As is deduced from immune-privileged organs such as eye and brain, cell-membrane bound surface molecules possibly protect against immune-mediated inflammation. Among them are the complement regulatory proteins, which inhibit the complement cascade, the presence of the Fas ligand to induce apoptosis of activated leucocytes, and the expression of MHC Ib, which block the killing activating receptor on NK cells and on cytotoxic lymphocytes (Ferguson et al. 2002; Niederkorn 2006). The CAM implantation assay can be used to coarsely check the status of immune privilege for organs in comparison. The mild inflammatory response of the CAM observed here after implantation of the heart (Fig. 3.1e, f) can be explained by the growing evidence that INIM responses and TLR signaling underlie myocardial adaptation to ischaemia (Chao 2009; Valeur and Valen 2009).

3.2.2 Follicular Atresia

Follicular atresia affects 99.9% of follicles in all stages of growth. Follicular atresia develops in two ways (Rodgers and Irving-Rodgers 2010; Spanel-Borowski 1981).

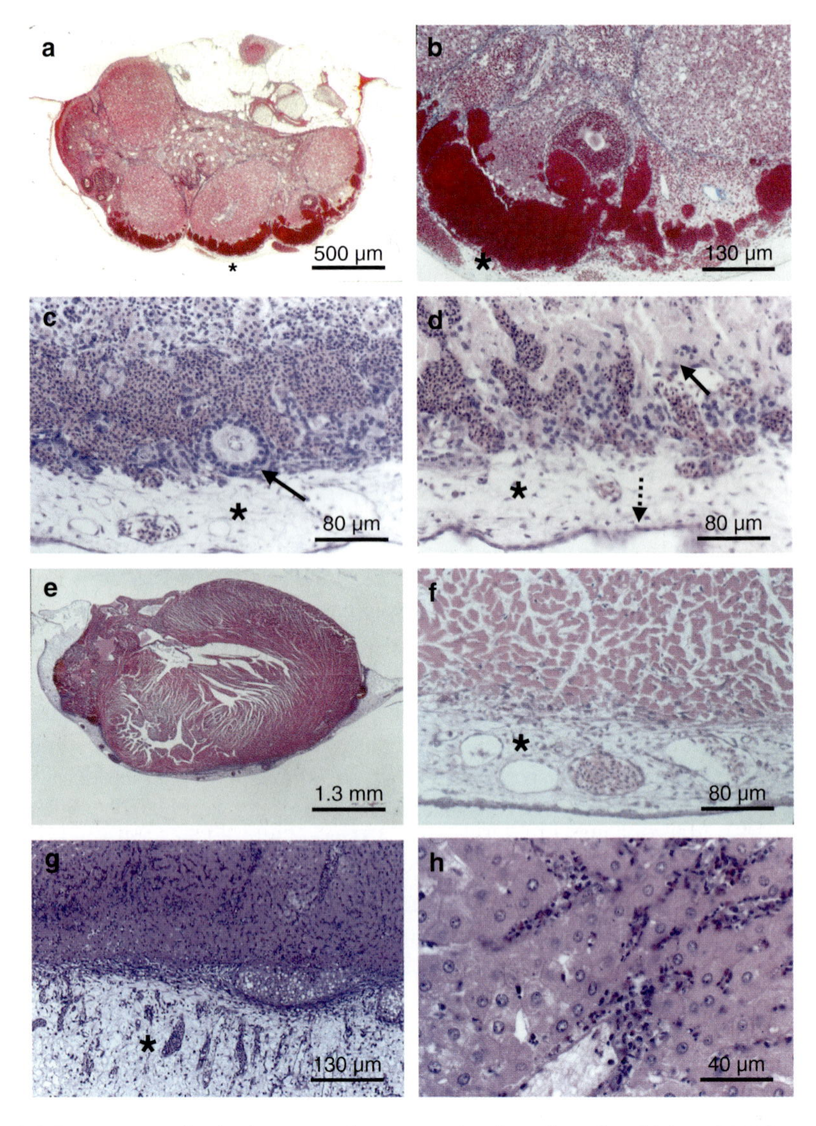

Fig. 3.1 Hamster ovaries in the stage of proestrus implanted on the chicken CAM between days 9 and 14 of incubation (**a**–**d**). The heart and a liver lobe were used in comparison (**e**–**h**). (**a**–**d**) The CAM mesenchyme (*asterisks*) lacks an inflammatory reaction against the ovarian implants. Vascularization is induced from the CAM towards the implant (**a** and **b**). Ingrowing vessels contain nucleated chicken erythrocytes (**c** and **d**). The primordial follicle looks healthy (*arrow* in **c**), while a rim of cortical tissue becomes necrotic (*arrow* in **d**). The chorionic epithelium is absent, and the allantoic epithelium is maintained (*broken arrow* in **d**). (**e** and **f**) The heart implant causes a mild inflammatory response in the mesenchyme (*asterisk* in **f**) with disappearance of the chorionic epithelium, prominent vessels and some leucocytes. (**g** and **h**) The liver implant leads to a heavy inflammatory response with dense infiltrates of leucocytes both in the mesenchyme (*asterisk* in **g**) and in the implant. All figures by HE. Adapted from Spanel-Borowski (1989)

In preantral follicles, pathway A affected the oocyte first (Fig. 3.2a–d). The zona pellucida ruptured, and the oocyte underwent necrosis and fragmentation, whereas granulosa cells remained intact. They appeared to transform into the interstitial cortical cells, whereby resumption of meiosis can occur. In antral follicles, pathway B displayed apoptotic cells in the granulosa cell layer, which gradually disappeared in a cystic follicle (Fig. 3.2e–h). It often contained a morphologically intact oocyte. Pathway B was associated with hypertrophic thecal cells. Early stages might show a few eosinophils in the thecal cell layer, while advanced stages depicted connective tissue ingrowth into the former lumen together with recruited macrophages and capillary sprouts. In other words, regressing preantral follicles do no harm, yet antral follicles provoke a local inflammatory response. Follicles might sense danger when nutritional and endocrine factors are being shortened at the onset of regression. The kind of protection against danger signals obviously differs when comparing preantral and antral follicles. It is possible that the regulator protein CD46 is involved. CD46, which is ubiquitously present, cleaves C3b into inhibitory C3b (see Sect. 3.1); the deposition of C3b is prevented as also is the complement-mediated cell attack in granulosa cells from preantral follicles. Dying granulosa cells of antral follicles could produce less CD46 than intact cells. The activation of C3 receptors on granulosa cells by low levels of C3 subunits might induce the inflammatory response in regressing antral follicles. The hypothesis that a mild activation of the complement system plays a role in follicular atresia extends the insights into the mechanism of follicular atresia. Atresia is presently explained by apoptosis of granulosa cells through the Fas–FasL signaling system (Craig et al. 2007; Hussein 2005; Matsuda-Minehata et al. 2006) or by other forms such as necrosis and cell-death autophagy (Rodgers and Irving-Rodgers 2010; Van Wenzel et al. 1999). Macrophages immigrate for garbage phagocytosis (Wu et al. 2004). Granulosa cell death partially explains how the loss of the whole follicle and the recovery of the damaged place are managed.

3.3 Moderate Danger by Preovulatory Follicle Rupture and Acute Inflammation with Eosinophils

At the rupture site of the preovulatory follicle, oedema forms because of an increase in the permeability of the capillaries, and leucocytes are recruited into the whole follicle wall releasing MMPs for the degradation of the basement membrane (Brännström et al. 1994b; Espey 1994). The fibrinolytic activity augments in the follicle wall through activation of plasminogen activator (Reich et al. 1985). The whole follicle wall collapses, and capillaries sprout from the vascular thecal cell layer into the avascular granulosa cell layer associated with leucocyte infiltrates between the luteinizing cells. Some steps of the periovulatory events are documented for the bovine preovulatory follicle in Fig. 3.3. It is noteworthy that toluidine blue-stained mast cells are absent in the wall of the preovulatory follicle

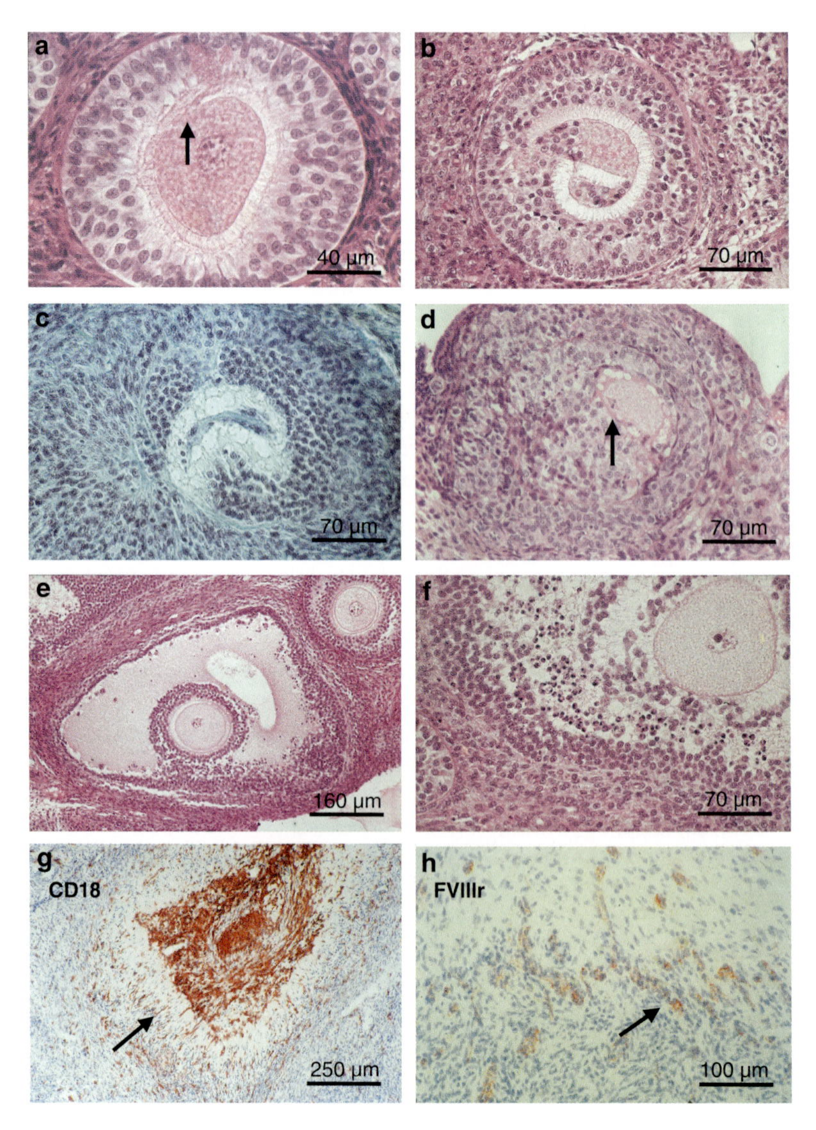

Fig. 3.2 Follicular atresia and two regression forms in ovaries from dogs (**a**–**c**, **e**, **f**), rats (**d**) and cows (**g** and **h**). (a–d) In pathway A, the oocyte is damaged and the zona pellucida has ruptured (*arrow* in **a**), whereas granulosa cells look healthy (**a** and **b**). Follicle cells of terminal atretic stages seem to transform into interstitial cortical cells (**c**), and the mitotic figure (*arrow*) points to the resumption of meiosis (**d**). (**e**–**h**) In pathway B, the oocyte appears intact, whereas granulosa cells undergo apoptosis (**e** and **f**). In terminal stages, CD18-positive macrophages accumulate, hyalinization of the connective tissue occurs as a sign of repair (*arrow* in **g**) as does the presence of capillary sprouts (*arrow* in **h**). (**a**, **b**, **d**–**f**) HE; (**c**) HOPA; (**e** and **f**) indirect immunohistology of paraffin sections for CD 18 and FVIIIr. Adapted from Spanel-Borowski (1981) and Spanel-Borowski et al. (1997)

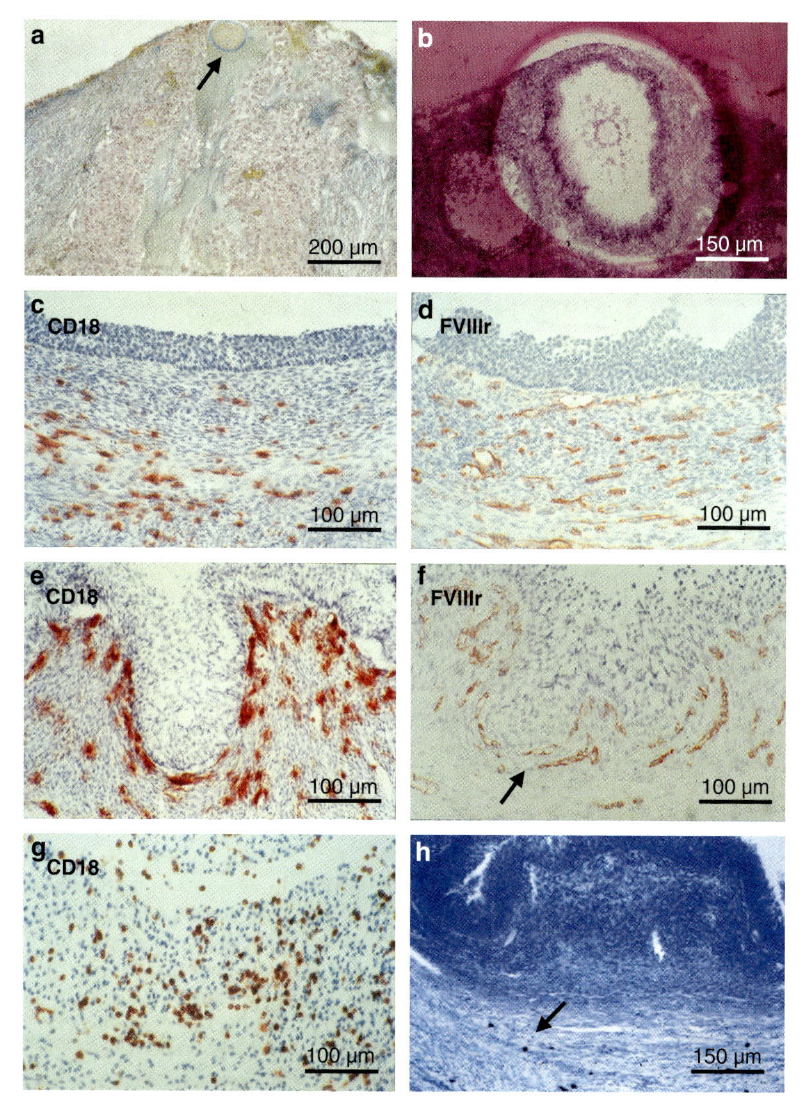

Fig. 3.3 Changes of the preovulatory follicle at rupture time in rabbit (**a**), golden hamster (**b**) and cow (**c**–**h**). (**a**) The oocyte is captured at the rupture side (*arrow*) during breakdown of the luteinizing follicle wall (HOPA). (**b**) The fibrinolytic activity is localized as a lytic area in the fibrin film on a cryostat section. (**c** and **d**) An antral follicle contains fewer CD18-positive cells in the theca than in the adjacent cortical tissue as shown with indirect immunohistology of paraffin sections (**c**). The CD18 antigen is a marker for granulocytes, monocytes and macrophages. Microvessels are responsive to the endothelial cell marker FVIIIr and equally distributed (**d**). (**e** and **f**) The preovulatory follicle shows a gyrated wall, accumulated CD18-positive cells in the zone of the basement membrane (**e**), and a conspicuous inner and outer microvascular bed (*arrow*, **f**). (**g**) The granulosa cell layer of the freshly ruptured follicle is invaded by CD18-positive cells. (**h**) Mast cells as stained with toluidine blue (*arrow*) are absent in the preovulatory follicle wall. Adapted from Spanel-Borowski (1987), Spanel-Borowski et al. (1996) and Reibiger and Spanel-Borowski (2000)

(Fig. 3.3h), because they are involved in functions of INIM and adaptive immunity (Kalesnikoff and Galli 2008). The rupture of a preovulatory follicle compares with a physiological wound, and the subsequent capillary sprouting with connective tissue ingrowth brings to mind a healing process. In the cyclic ovary, the lesions of follicle ruptures are extremely well compensated, because the inflammatory events do not turn into pathological reactions. We here suggest that the effective control of tissue integrity depends on the surveillance function of INIM with granulocytes as executives and with a novel type of epithelial effector cells (see below). Thus, the current concept that ovulation is an acute inflammation is confirmed and extended in such a way that the reaction is part of ovarian INIM responses to danger signals. Our concept is supported by gene profile analysis of preovulatory follicles revealing granulosa cells with genes for acute inflammatory action and for immunosurveillance (Richards et al. 2002, 2008). The dominant leucocytes are identified as macrophages and segmented granulocytes (Wu et al. 2004). The latter are described as neutrophils in preovulatory follicles of rats and humans (Brännström and Enskog 2002), and as eosinophils in sheep and pigs (Murdoch and Steadman 1991; Standaert et al. 1991). We localized eosinophils with Sirius red staining and $CD18^{+}$ cells with immunohistology in subsequent sections of bovine and human ovaries (Aust et al. 2000; Reibiger and Spanel-Borowski 2000; Rohm et al. 2002). At the early developmental stage of bovine tissue, eosinophils accumulated in CL septa and around the thecal microvessels indicating their control in the recruitment of eosinophils (Fig. 3.4a–d). The number of eosinophils reached approximately 90% of the $CD18^{+}$ leucocyte pool, and they populated the former thecal cell layer at a higher density than the luteinizing granulosa cells (Fig. 3.4e). In our view, eosinophils are representatives for the late phase of the periovulatory period when, after follicle rupture, the collapsed follicle transforms into a CL.

The traditional function of eosinophils is related to cell and tissue damage after parasite infections and the release of cytotoxic granule-associated proteins. Eosinophils appear to do much more because of their novel roles in immune regulation, in angiogenesis and in tissue repair (Munitz and Levi-Schaffer 2004; Blanchard and Rothenberg 2009). Eosinophilic cytokines (interleukins, TNF-α, IFN-γ) affect T-cell proliferation and antigen presentation, and activate mast cells through the major basic protein release in an IgE-independent way. Eosinophils contain angiogenic factors such as VEGF and b-FGF and also synthesize pro-angiogenic cytokines such as IL-6 and Il-8, and granulocyte macrophage colony-stimulating factor (GM-CSF). The transforming growth factor β (TGF-β) with fibrinogenic capacity and the eosinophilic cationic protein influences fibroblast proliferation, collagen synthesis and inhibition of proteoglycan degradation. The extracellular matrix is modulated by expression of eosinophilic MMP-9 and of MMP inhibitors. Eosinophils contain neuromediators such as neurotrophic growth factor (NGF) and substance P (SP). Altogether, eosinophils are not innocent bystanders but active contributors to inflammatory processes of broncho-pulmonary and gastro-intestinal disorders by contributing to immune regulation, angiogenesis and tissue remodeling (Nissim

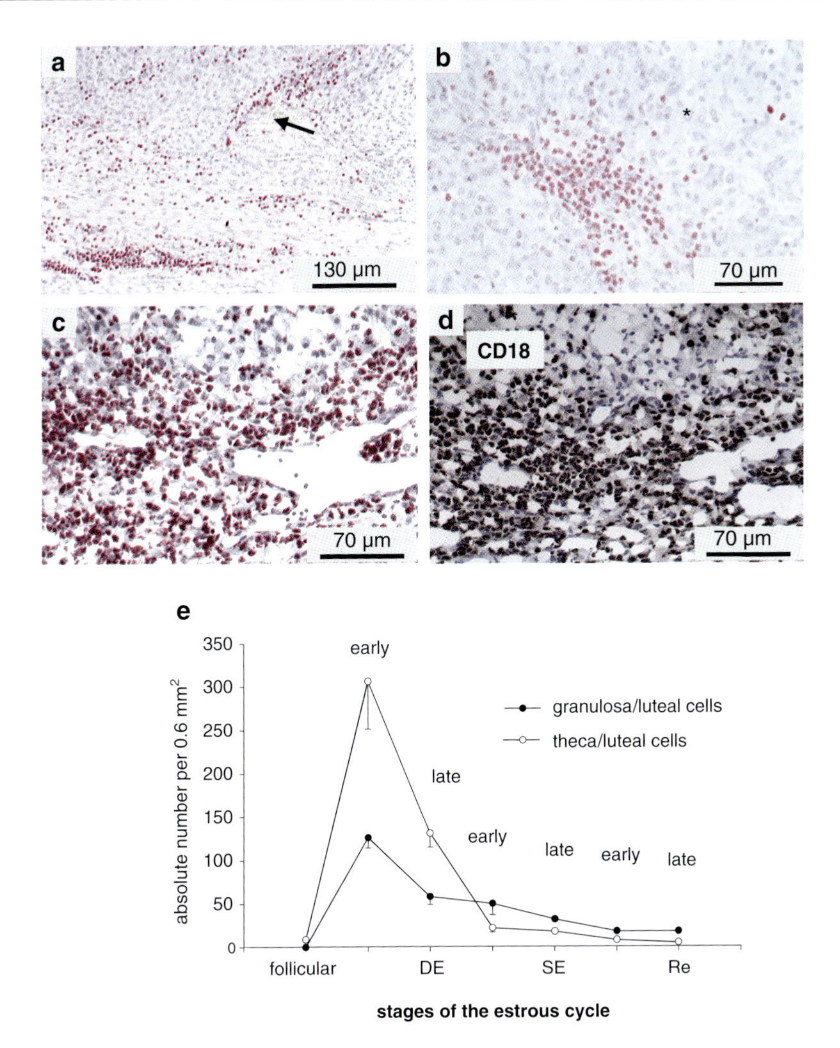

Fig. 3.4 Numerous eosinophils in the former thecal layer of the developing CL. (**a** and **b**) Sirius red staining shows eosinophils in the septum (*arrow*) and the very periphery of the CL (**a**). Fewer eosinophils are seen between luteinizing granulosa cells (*asterisk*) than the former thecal cell layer (**b**). (**c**–**e**) Most of the eosinophils, which are associated with a post-capillary venule (**c**), represent CD18-positive cells in the immunostained section (**d**). In comparison with CD18-positive cells, 90% of eosinophils are counted at the stage of early development, 54% at the late stage and 10% at the stage of secretion (**e**). Adapted from Reibiger and Spanel-Borowski (2000) and Rohm et al. (2002)

Ben Efraim and Levi-Schaffer 2008). Likewise, eosinophils might play a comparable role during transformation of a ruptured follicle into a CL. The mechanism of eosinophil recruitment, which follows a general principle for every tissue, is likely adapted to the needs of the follicle in transition.

3.3.1
Recruitment of Eosinophils and SP-Like Expression

Eosinophils are derived from the myeloid lineage in the bone marrow, and mature eosinophils are released under Il-5 control into the blood. Eosinophil emigration from the blood to the tissue depends on main chemoattractants such as eotaxin, RANTES (regulated on activation, normal T cell expressed and secreted), and the C3a and C5 anaphylatoxins, their receptors included (Blanchard and Rothenberg 2009; DiScipio and Schraufstatter 2007). They are able to reinforce molecules for the stages of emigration. For rolling, the eosinophil employs L-selectin (CD62L/CD162) and the counter-receptor P selectin (CD62P) on endothelial cells; for firm adhesion, integrins ($\alpha_4\beta_1$, CD49d/CD29) are responsible on eosinophils, whereas intercellular and vascular adhesion molecules (ICAM, VCAM-1) are present on endothelial cells. For diapedesis, eosinophils homotypically interact with platelet endothelial cell adhesion molecules (CD31) at the lateral endothelial cell membrane and with the C3a/C5a receptor complex by evoking a stable retraction of endothelial cells (DiScipio and Schraufstatter 2007). The extracellular matrix is exposed for connective tissue penetration by activating complement receptors among other factors. When C3a and C5a bind to eosinophil receptors the MMP-9 is released and matrix components are degraded.

Eosinophils are prominent in the mucosa of the intact intestinal tract. The early CL development is largely unknown as a tissue of conspicuous eosinophil recruitment (Aust et al. 2000; Reibiger and Spanel-Borowski 2000). This finding correlated with a CD62P upregulation in endothelial cells from previous thecal microvessels (Rohm et al. 2002). Simultaneously, eosinophils with CD62L immunopositivity assembled around the responsive vessels, thus confirming rolling of eosinophils by selectin interaction in the bovine tissue. The firm adhesion, which depends on integrin interaction with VCAM, cannot be investigated in the bovine ovary, because none of the available human monoclonal antibodies selectively detects the CD29a/CD49 molecule on eosinophils. In leucocyte-depleted granulosa cell cultures derived from human preovulatory follicles, RANTES, not eotaxin, is the major eosinophil-attracting chemokine at 12 h of TNF-α application (Aust et al. 2000).

We have several reasons to assume that neuropeptides, in particular SP-like molecules, are also involved in specifying the role of eosinophils in the early CL. Substance P belongs to the tachykinin family with members such as neurokinin A and B (NKA, NKB). They share a carboxy-terminal sequence of five amino acids essential for receptor interaction, whereas the amino-terminal sequence of 5–7 residues provides the subtype selectivity. Two separate genes encode precursor proteins called pre-protachykinin A and B (PPT-I, PPT-II), which are chopped into small peptides by post-translational enzymatic processing (Severini et al. 2002). The PPT-I is cleaved into SP and NKA, and the PPT-II into NKB. Substance P and NKA prefer the neurokinin-1 receptor (NK-1R), NKA preferentially binds to NK-2R, and NKB to NK-3R. Substance P, which is secreted by nerves as well as by inflammatory cells such as eosinophils and DCs, exerts proinflammatory effects through its NK-1R by release of inflammatory mediators. Smooth muscle cells relax, vascular permeability increases, adhesion molecules for leucocyte recruitment are upregulated, and functions change in immune cells and in epithelial cells. Elevated levels of SP, NK-1R and eosinophil accumulation characterize the neurogenic inflammation in allergic lung disorders, chronic bowel disorders and rheumatoid arthritis (O'Connor et al. 2004; Lambrecht 2001).

The appearance of eosinophils in the developing CL coincided with the mRNA presence of SP and of NK-1R (Fig. 3.5; Reibiger and Spanel-Borowski 2000; Reibiger et al. 2001). The reverse transcription polymerase chain reaction (RT-PCR) and nested PCR technique showed signals for the transcripts of the PPT-I and PPT-II genes as well as of the NK-1R and NK-3R at all CL stages (Fig. 3.5; Brylla et al. 2005). The NK-2R transcript was missing. By immunohistology, SP-positive networks of fiber-like structure were seen in the former thecal cell layer, the septum and also between luteinizing cells of the developing CL (Fig. 3.6a, b; Reibiger et al. 2001). The SP-positive immunoreactive network appeared to disintegrate in the CL stages of secretion and regression. The dot blot analysis for SP excluded antiserum-related cross-reactivities with NKA and NKB, and the semiquantitative analysis revealed a trend towards a lower immunoreactivity at CL stages of secretion and regression compared to the developmental stage (Fig. 3.6e, f). No staining pattern was observed with a normal rabbit serum, yet preincubation of the SP antiserum with the SP peptide did not inhibit the immunoresponse in the developing CL (Fig. 3.6c, d). For this reason, the positive staining response is judged to be an SP-like answer. Cross-reactivity between the antiserum against SP and a still undefined tachykinin member may have occurred. The tachykinin family comprises more than 40 members, and members are growing as is deduced from novel endokinins (Page et al. 2003; Severini et al. 2002). They arise from the PPT-4 gene, which is spliced into four alternative transcripts which among other endokinins generate hemokinin-1. Its mRNA is detected in leucocyte subtypes, eosinophils included (Klassert et al. 2008). Hemokinin-1 preferentially binds to the NK-1R. Of note, ovaries from adult NK-1R gene-deficient mice contained an increased number of luteinized and unruptured follicles (Löffler et al. 2004b). Endothelial cell cultures and macrophages, both derived from bovine CL, showed Ca^{2+} influx under agonist treatment of NK-1R as is verified by calcium imaging (Brylla et al. 2005). These diverse findings point to the functionality of the NK-1R in the ovary and thus to the interaction with a specific tachykinin ligand. An undercover member with SP-like structure and preferential binding to the NK-1R could have been detected by immunohistology and by dot blot analysis in the bovine CL. It is assumed that the still concealed member is expressed by non-neuronal cells and that a tachykinin-receptor-guided co-regulation leads to a selective immune performance. The acute inflammatory response is the outcome in periovulatory events. Debeljuk (2005) has collected evidence that tachykinins are synthesized in granulosa and luteal cells to act as intra-ovarian modulators. It is

a

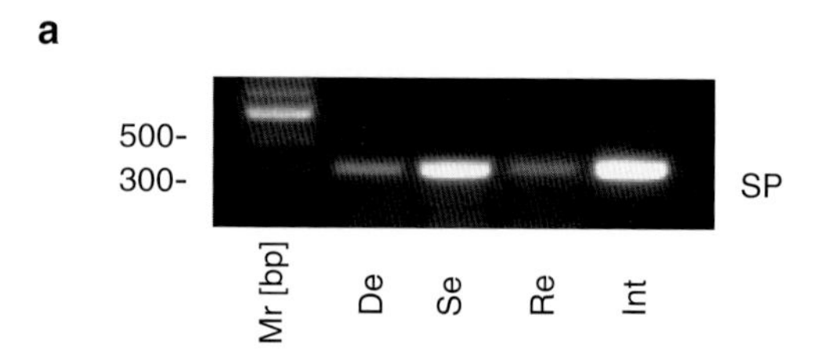

b

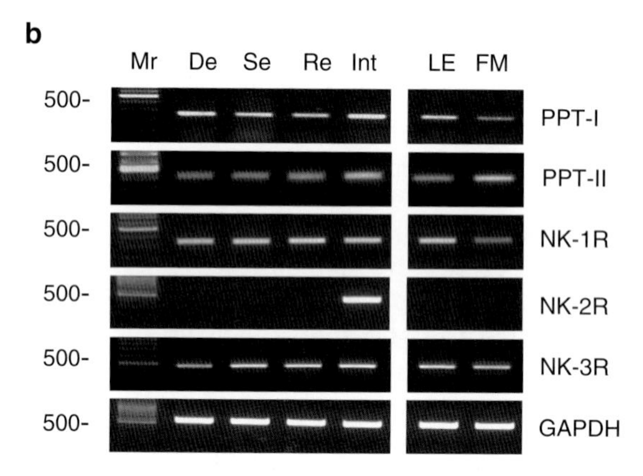

Fig. 3.5 The mRNA presence of SP and of tachykinin precursors (preprotachykinin gene I/II, PP-I, PP-II) together with the neurokinin-1–3 receptor (NK-1R, NK-2R, NK-3R) in bovine CL. The RT-PCR and nested RT-PCR analyses were performed with RNA isolated at various stages (*De* development, *Se* secretion, *Re* regression) as well as from intestine (*Int*), luteal endothelial cells (*LE*) and from follicle macrophages (*FM*). Comparable amounts of amplified 566 bp GAPDH fragment were used as internal control, and no sample DNA was applied as negative control (not shown). The results are representative of four independent experiments. *Mr* mass ruler™ DNA ladder (100-bp ladder). Adapted from Brylla et al. (2005)

noteworthy that neuroimmunology focuses on the role of neuropeptides as immunomodulators of DCs. Neuropeptides might mediate DC migration (Dunzendorfer and Wiedermann 2001). Any progress on tachykinin specification in the early CL is hampered by the lack of monoclonal antibodies against the small tachykinin molecules deviating in selectivity by just a few amino acids. Furthermore, the availability of specific agonists and antagonists is limited, and intraperitoneal application is impeded by the unspecific inflammation of the peritoneum and of the periovarian capsule (Löffler et al. 2004a). Ovarian eosinophils are not available in sufficient numbers for immunological studies. In other words, the steps showing how tachykinin-receptor-dependent actions interact with eosinophils in the developing CL will have to wait for their final disclosure.

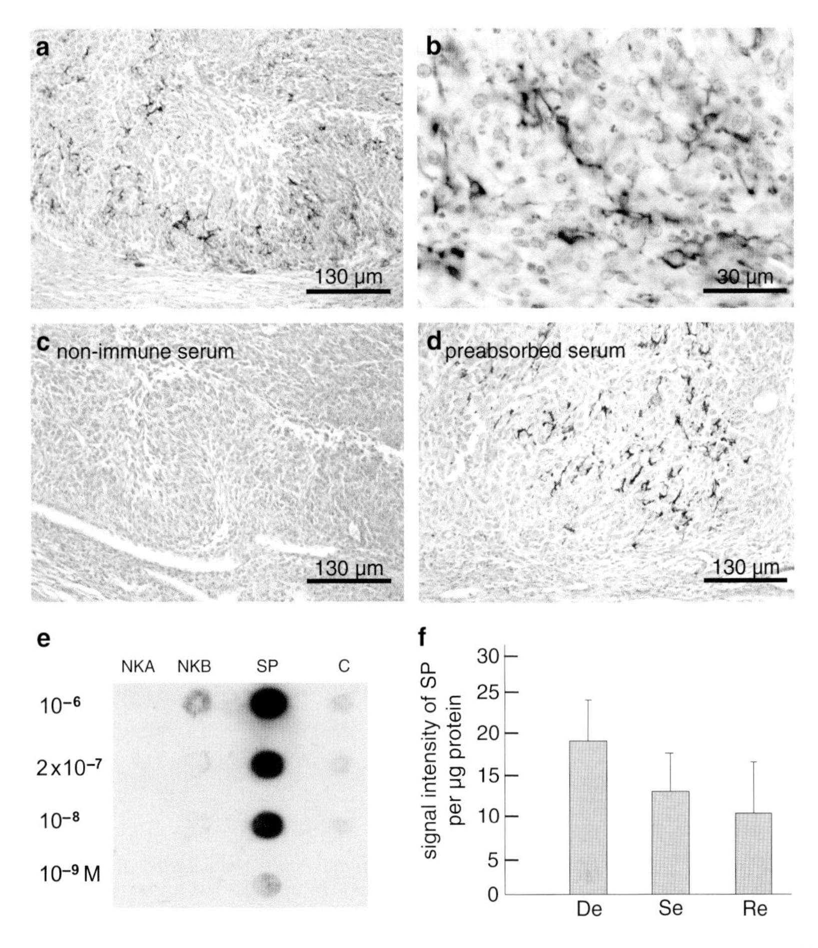

Fig. 3.6 Substance P-like immunoresponse in the developing bovine CL. (**a–d**) The fiber-like network is noted in the periphery as revealed by indirect immunohistology in paraffin sections and the use of a polyclonal rabbit antiserum (**a**). The response appears between luteal cells (**b**). The control with a non-immune antiserum is negative (**c**), yet remains positive by use of a preabsorbed antiserum. The method is indicated: 5 μl of serially diluted SP (1–10^{-6} μM; Bachem, Bubendorf, Switzerland) was incubated with 100 μl of the 1:1,000 diluted SP antiserum at 4°C overnight, and centrifuged twice at 1,200 rpm for 10 min. Preabsorption control with bovine intestine sections lacked SP-positive nerve fibers (not shown). (**e** and **f**) The selectivity of the antiserum against SP is validated with the dot blot technique for CL proteins immunoblotted with neurokinin A and B (NKA, NKB) in (**e**). The semiquantitative evaluation of the signal intensities show no statistically significant changes between the estrous cycle stages (**f**). Adapted from Reibiger et al. (2001)

3.4

Severe Danger in Superovulated Ovaries with Intra-ovarian Oocyte Release (IOR) and Thrombus Formation

Superovulations by gonadotrophin treatment of immature, prepuberal rats and mice are useful to synchronize the ovarian cycle and to catch the time-sequence of follicle rupture. Yet, it is widely observed that the pharmacological stimulation is associated with tissue damage, which heavily surmounts that of a physiological follicle rupture. In superovulated ovaries, the preovulatory follicles lose the orientation towards the surface epithelium and release the oocyte into the interstitial cortical space. The process of lost orientation is termed IOR. Already, small preantral and antral follicles show IOR (Spanel-Borowski et al. 1982; Spanel-Borowski and Aumüller 1985). It was incomplete with the retention of the oocyte in the follicle compartment and complete when the oocyte is expulsed (Fig. 3.7a, b). Intra-ovarian oocyte release was observed in immature and in adult laboratory animals in diestrus (mice, rats, Syrian hamsters, white-footed mice), which signifies that intra-ovarian factors ultimately control follicle rupture. The factors also appear to be activated during follicular atresia, because IOR occurred together with deformation of the oocyte, apoptotic granulosa cells and an irregular granulosa cell layer. In superovulated ovaries of rats and golden hamsters, severe tissue damage developed because of increased IOR (Spanel-Borowski et al. 1983b). The oocyte was seen in a hemorrhagic lake at the rupture site, and appeared with the mitotic spindle as a sign of resumed mitosis in the cortical tissue or even in microvessels (Fig. 3.7c–e). The number of IOR was roughly three times higher in the group with the high gonadotrophin dose at 24 h after hCG application compared to the low-dose group (Table 3.1; see Fig. 2.1a). Furthermore, because many IOR were noted at 48 h of hCG/LH application, ovulation time was extended for 12 h in the high gonadotrophin group (Table 3.1). The IOR was associated with prominent thrombus formation in large microvessels of the medulla, from there reaching into the cortical region (Fig. 3.7f). The higher the gonadotrophic dose that had been given, the stronger was the thrombus formation. Thrombi were fresh on days 3 and 4 of PMSG stimulation, whereas they were organized on subsequent days 5–7 (Spanel-Borowski and Heiss 1986). Of note, when the fibrinolytic activity was assessed with fibrin-coated cryostat sections, the activity decrease inversely correlated with thrombus occurrence (Fig. 3.7g). In spite of the severe IOR-dependent damage, superovulated ovaries quickly recuperated for the next ovarian cycle. They thus represent an excellent animal model to reveal the molecular pattern of INIM action from wounding to self-repair. Circumspection should be given to repeatedly stimulated ovaries which depict a low number of CL and of IOR as signs of hypoovulations, probably due to an insufficiently developed capillary bed in the thecal cell layer (Löseke and Spanel-Borowski 1996).

The IOR with thrombus formation is not species-specific, because the findings were confirmed in adult rabbits with and without ovarian stimulation (Fig. 3.8a–d).

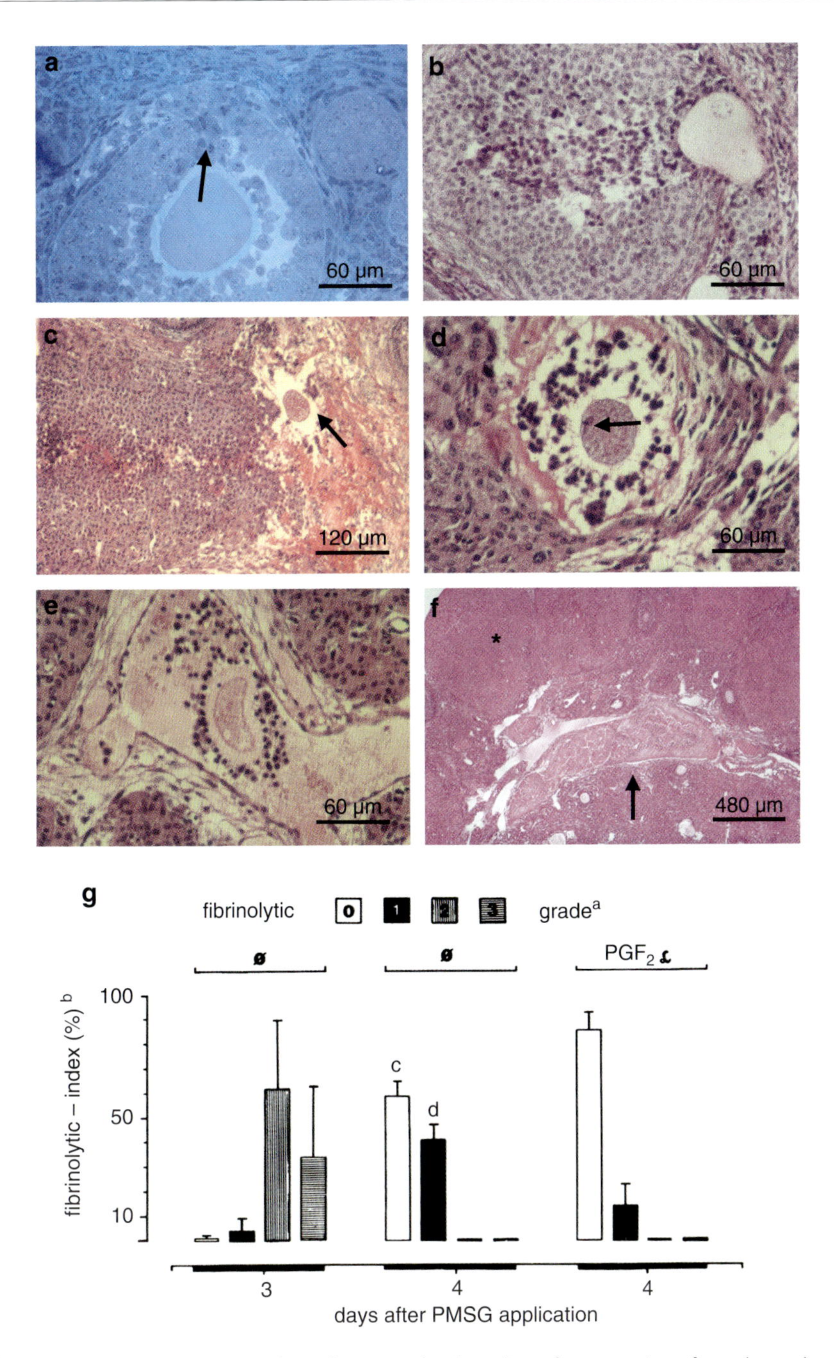

Fig. 3.7 Intra-ovarian oocyte release in HE-stained sections from ovaries of rats (**a**, **c**–**e**) and golden hamsters (**b** and **f**). (**a** and **b**) The preantral follicle of small size from a 21-day-old rat (*arrow* in **a**) and the large one from a cyclic hamster ovary show incomplete IOR (**b**).

Table 3.1 Superovulated 28-day-old golden hamster[a] with IOR[a] and thrombi[a]

Dose	Hours after hCG			
	12	24	36	48
Absolute number of IOR[b]				
10 IU PMSG 10 IU hCG	**2.3** ± 1.5	**2.8** ± 1.7	**1.7** ± 1.4	None
50 IU PMSG 25 IU hCG	**5.6** ± 3.6	**8.4** ± 3.0	**7.6** ± 4.4	**6.6** ± 3.8
Thrombi[c] *in the left/right ovary*				
10 IU PMSG 10 IU hCG	**0/9**	**9/2**	**3/10**	None
50 IU PMSG 25 IU hCG	**0/9**	**5/2**	**7/0**	**6/3**

Data are means ± SD from at least five animals per group
[a]IOR for intra-ovarian oocyte release in completely sectioned ovaries
[b]IOR for thrombus occurrence in completely sectioned ovaries
[c]Scheme for gonadotropin injection is given in Fig. 2.1a

When 3-month-old rabbits had been stimulated with FSH and ovulation induced with hCG, many IOR from preovulatory follicles occurred (Spanel-Borowski et al 1986). Remnants of the follicle wall and of the cumulus complex floated in small and large oedematous lakes. Thrombi were observed in adjacent blood and lymph vessels sometimes enclosing granulosa cells and oocytes. To validate the influence of prostaglandins on IOR occurrence, lonazolac and indomethacin, both inhibitors of the arachidonic acid cascade, were applied into the vagina at the time of hCG injection. The IOR ruptures increased from around 5 IOR per ovary at 15 h of hCG application to 20 IOR at the 28 and 48 h time points, independent of the

Fig. 3.7 (continued) (**c** and **d**) A complete IOR with follicle transformation into a CL is noted in superovulated ovaries from a 27-day-old rat on day 3 after PMSG. The expulsed oocyte is seen in a hemorrhagic lake (*arrow* in **c**), with cumulus expansion and mitotic figure of resumed meiosis in the interstitial cortical tissue (*arrow* in **d**) as well as in a venule (**e**). (**f**) A fresh thrombus occurs in a vein of the ovarian medulla (*arrow*) and belongs to a superovulated ovary with many CL (*asterisk*) from a 28-day-old hamster on day 4 after PMSG stimulation. (**g**) The thrombus presence coincides with the decrease in fibrinolytic activity on day 4 after PMSG stimulation of superovulated hamster ovaries as determined by the fibrin slide technique on cryostat sections. (*a*) Four grades were classified: grade 0, no lytic areas; grade 1, lytic areas <2 mm in diameter 3 h after incubation (see Fig. 3.2b); grade 2, lytic areas >2 mm in diameter; grade 3, lytic and confluent areas 1 h after incubation. The decline in fibrinolytic activity on day 4 after PMSG is enhanced after treatment with prostaglandin $F2_{\alpha}$. (s.c. 500 μg PG $F2_{\alpha}$ at 08:00 hours of day 4). (*b*) The fibrinolytic index compares with the percentage of section sets from a completely cut ovary with different fibrinolytic grades. (*c* and *d*) $p < 0.001$ versus grade 0/1 of days 3 and 4 (with $PGF2_{\alpha}$) after PMSG stimulation. Adapted from Spanel-Borowski and Aumüller (1985), Spanel-Borowski and Heiss (1986) and Spanel-Borowski et al. (1982, 1983)

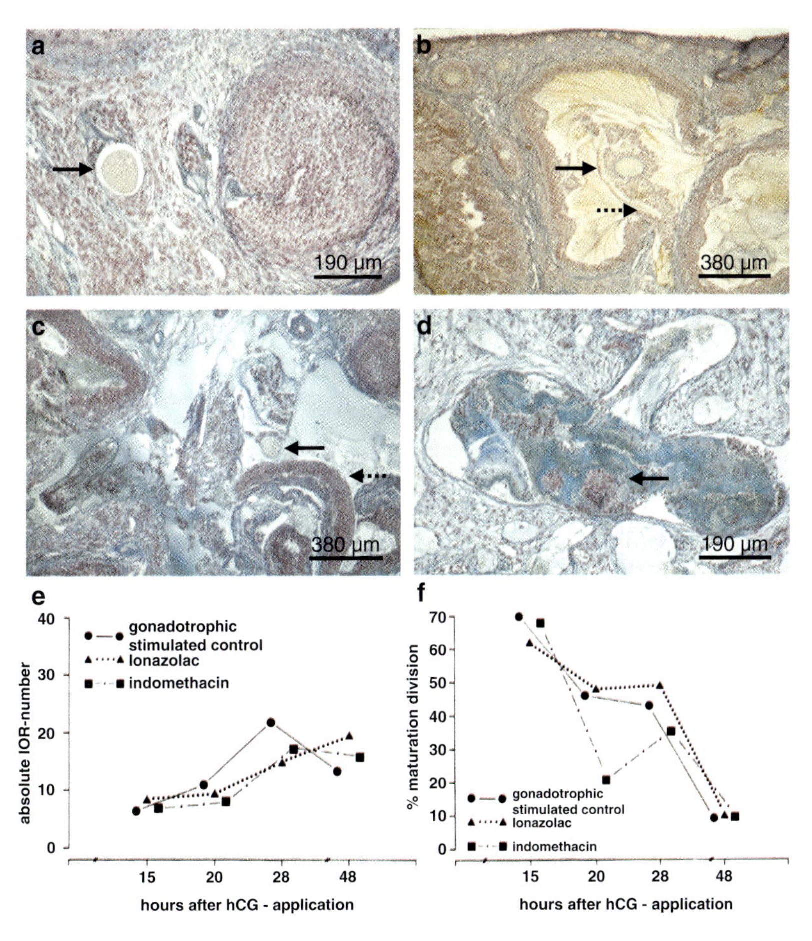

Fig. 3.8 Intra-ovarian oocyte release in superovulated ovaries from 3-month-old rabbits is studied 20 h after hCG injection in HOPA-stained paraffin sections (**a–d**). (**a**) The control section displays a complete IOR in a collapsed antral follicle with the oocyte in the cortical tissue (*arrow*). (**b**) The oocyte of an incomplete IOR (*arrow*) is seen in the antrum. Associated granulosa cells extend in the rupture site (*broken arrow*). (**c**) The complete IOR has heavily traumatised the cortical tissue. The oocyte (*arrow*) as well as pieces of the luteinized follicle wall (*broken arrow*) and of the cortical tissue floats in an oedema. (**d**) A fresh thrombus with granulosa cell complexes (*arrow*) appears in a vein. (**e** and **f**) Superovulated rabbits were additionally treated with lonazolac or indomethacin, both enzyme inhibitors of the arachidonic acid cascade. At the time of ovulation induction with i.v. injection of 100 IU hCG, 1 mg inhibitor in 0.1 ml 0.9% NaCl per kg body weight was introduced into the vagina. Type 5 follicles above 600 µm in diameter and cysts were counted in completely cut ovaries, oocytes with meiotic signs were recorded and the percentage calculated. Because the data are the mean from $n = 3$ ovaries, statistics were not possible. Adapted from Spanel-Borowski et al. (1986)

inhibitor treatment (Fig. 3.8e). Oocytes with signs of resumed meiosis had already decreased at the 15-h time point in the three groups (Fig. 3.8f). Thus, IOR extended over a longer period after ovulation induction in contrast to maturation division. The intensity of thrombus formation was high for all treated groups at 28 h of hCG treatment, i.e. it correlated with the IOR peak.

Collectively, the powerful IOR-dependent danger signals generated by severe tissue damage likely cause thrombus formation through activation of the coagulation system. This connexion strongly supports the suggestion that danger signals influence the complement system, which cross-react with the coagulation system through the C1q–C1 receptor complex on endothelial cells and platelets (Köhl 2006b). Also, the plasma-derived factor XII, known as Hageman factor, that binds to components of the extracellular matrix, is activated and generates the coagulative cascade among others (Medzhitov 2008). Thrombus formation reflects INIM defence and is an element of the prominent acute inflammatory response. It comprises increased vascular permeability, the generation of chemotactic factors for the recruitment of leucocytes, and, finally, tissue recovery. None of the severe IOR-related tissue lesions leave permanent changes. The next ovarian cycle runs regularly, and no morphological traces point to the recent catastrophe of IOR lesions. This differs from the healing process of a wound, which produces granulation tissue finally becoming a scar. In line with the hypothesis that INIM has the potential to regulate the ovarian repair process most efficiently, major effectors should be defined. In our opinion, CK^+ cells have the capacity to be recognized as DCs.

Chapter 4
Cytokeratin-Positive Cells (CK^+) as Potential Dendritic Cells

The skin and the mucosa represent anatomical barriers against invaders from the atmosphere or from the fluid-rich environment as the first line of defence (Turvey and Broide 2010). The barriers deliver mechanical and chemical protection by cell–cell coherence, and by secretion of unspecific anti-pathogenic agents. Immunological protection comes from nonlymphoid DCs with residency in the skin (Langerhans cells) and in the mucosa (Geissmann et al. 2010). They seem to have their own resident progenitors (Chorro et al. 2009). Lymphoid and plasmacytoid DCs in the skin and mucosa are renewed from bone marrow-derived precursor cells (Mellman and Steinman 2001; Merad and Manz 2009). The classical DCs are sentinels sensing antigen-presenting sites in the immature status and migrating as mature DCs to regional lymph nodes for the contact with lymphocytes. The DC population is difficult to isolate because they represent a minor population and because of poor expression of specific antigens (Banchereau and Steinman 1998). The mural granulosa cells of antral and preovulatory follicles can be considered as an anatomical barrier against the follicular fluid, and theoretically contain protective cells against dangerous invaders/danger signals. We have located CK^+ granulosa cells in antral and preovulatory follicles as well as in CL (human, cow). The origin of CK^+ cells is traced back to the sex cords of fetal ovaries and to primordial and primary follicles. The transient disappearance of CK^+ follicle cells is noted in preantral and small-sized antral follicles in the adult ovary (Löffler et al. 2000).

4.1
Dendritic Cells, the TLR System in General and in the Ovary

Dendritic cells are antigen-presenting cells and highly important in immunological defense against malignant tumours (Banchereau and Steinman 1998; Chan and Housseau 2008; Mellman and Steinman 2001). Dendritic cells represent a heterogeneous population by anatomical locations and specific missions, yet they are not characterized by rigorously defined antigens. Immature DCs of the lymphoid and plasmacytoid type from the bone marrow show positivity for the tyrosine kinase KIT receptor (KIT), and for intracellular MHC II-positive granules.

Immature DCs develop actin cables, and handle antigen processing. Mature DCs display no actin cables, yet show high surface membrane MHC II products. Dendritic cells express E-cadherin for homing after migration into regional lymph nodes, where DCs stimulate T cells, and prime naïve T cells, being partners of adaptive immunity (Iwasaki and Medzhitov 2010; Merad and Manz 2009). Mature DCs produce more Fc receptors for unspecific antigen uptake than immature DCs. Dendritic cells have the capacity to extend cell processes through tight junctions of gut epithelia, whereby DCs form tight junction proteins such as claudin and occludin to maintain the integrity of the epithelial barrier (Rescigno et al. 2001). The capacity to generate tight junction molecules could also be relevant in other organs when DC function is effective in microcompartments. Dendritic cells in the genito-urinary tracts are regulated in number and function by sex hormones and considered to be the first protection against invading pathogens (Iijima et al. 2008).

Dendritic cells sense exogenous and endogenous danger signals (see Sect. 1.1) with the help of the TLR family (Chan and Housseau 2008; Steinman and Banchereau 2007). The TLRs belong to the evolutionarily ancient PRRS (Hoshino and Kaisho 2008; O'Neill 2008; Takeda and Akira 2005; Takeuchi and Akira 2010). The TLRs constitute pleiotropic, yet tightly regulated, gateways for modulation of many thousands of genes. Each of the 12 TLR members is characterized by an extracellular leucine-rich repeat, forming a pocket for ligand binding, and a cytoplasmic toll-IL-1 receptor (TIR) domain for initiation of signal transduction after receptor dimerization (Takeuchi and Akira 2010). The different utilization of five TIR domain-containing adaptor molecules (Myd88, TRIF, TIRAP, TRAM, SARM) that are expressed either at the cytosolic site of the surface membrane or on endosomes provides selectivity of individual TLR-mediated pathways, and also guarantees controlled TLR signaling. The Myd88 (myeloid differentiation factor 88) is the key adaptor protein of two Myd88-dependent pathways, either activating the transcription factor activator protein-1 (AP-1) in control of cell growth, differentiation and cell death or the nuclear factor-κB (NF-κB). It opens the inflammatory and immunoresponsive repertoire comprising TNF-α, Il-1β, Il-6, Il 12, TGF-β, RANTES, MMPs, and prostaglandins to name a few. There also exists a Myd88-independent pathway termed TRIF (TIR-domain-containing adaptor protein inducing IFN-β)-regulated signaling for the generation of interferon-regulatory factor 3 (IRF3). The IRF-regulated genes lead to the production of anti-inflammatory IFN-α/β, which induce the activation of hundreds of IFN-controlled genes through the IFN receptor on the cell surface. Many of the factors belonging to the universe of TLR signaling play a role in the inflammatory process of follicle rupture (Fig. 3.3) and the transformation into a CL as a repair action (Craig et al. 2007; Stocco et al. 2007; Oktem and Oktay 2008). The source of intra-ovarian factors is attributed to immigrated leucocytes, to thecal cells and to granulosa cells. Granulosa cells with and without the expression of CK (Löffler et al. 2000; Serke et al. 2009, 2010) might produce different combinations of intraovarian factors.

The TLRs have favorite ligands of microbial molecules: TLR1 and 2 for lipoproteins, TLR4 for LPS, and TLR5 and 6 for zymosan and flagellin, respectively (Hoshino and Kaisho 2008; O'Neill 2008; Takeda and Akira 2005; Takeuchi and Akira 2010). These receptors are expressed at the cell surface, while TLR3 and TLR7-9 are found on endosomes and bind nucleic acids. The TLR4-mediated responses depend on signaling by the Myd88 pathway resulting in the activated transcription factors such as AP-1 (cell growth, cell differentiation, cell death) and in NF-κB (inflammatory and immunoregulatory responses). The TRIF pathway requires transfer of the TLR–ligand complex to endosomes for the activation of IRF-regulated genes for anti-inflammatory responses. The TLR4 is functionally active in CK^+ granulosa cells (Serke et al. 2009, 2010).

The earliest events in the TLR-derived signaling require co-regulatory receptors already shown for CD14 in regard to TLR4 activation (He et al. 2010; Miller et al. 2003a). In macrophages, CD36 is the co-receptor to recognize oxLDL and serum amyloid, which subsequently facilitates the assembly of TLR4 and 6 into a heterodimer (Stewart et al. 2010). It signifies that CD 36 is proximal to danger sensing by the TLR heterodimer. According to Stewart et al. (2010), the heterodimeric complex then elicits the TRIF-dependent and MyD88-dependent signaling pathways finally producing the chemokine RANTES or ROS, respectively. RANTES might support eosinophil accumulation in CL of an early stage (Fig. 3.4), while ROS by acting as second messenger of signal transduction in INIM (Kohchi et al. 2009) could inhibit the inflammatory NF-κB pathway. It is assumed that endogenous danger ligands are processed to induce genes for tissue inflammation and repair. The presence of varying co-receptors appears to contain the secret of how the complex TLR signaling pathways are delicately tuned to mediate a physiological inflammation (Medzhitov 2008, 2010a), to which the periovulatory event compares (Fig. 3.3). Mast cells, which are seen in the cortical tissue and medullary stroma, not in the periovulatory follicle wall (Fig. 3.3h; Reibiger and Spanel-Borowski 2000), might inhibit the recruitment, differentiation and activation of T cells, and thus diminish the connection between INIM and adaptive immunity. The speculation is in line with new insights in mast cell biology showing that mast cells could be mediators of adaptive immunity (Kalesnikoff and Galli 2008; Wasiuk et al. 2009).

In addition to a cell membrane co-receptor, which is required for a functional heterodimeric TLR complex, cross-talks should be discussed between TLR and complement receptor signaling pathways downstream of signal initiation. Such a transactivation between two intracellular pathways might modulate the final signaling signature (Hajishengallis and Lambris 2010; Hawlisch and Köhl 2006). The same might hold true for the NK-1R-tachykinin system, because they appear to be influential in inflammatory diseases of the respiratory, gastro-intestinal and musculoskeletal systems (O'Connor et al. 2004). Of note, there is recent evidence for co-regulation of NK-1R and TLR4 gene expression in airway cells from children with airway bacterial colonization (Grissell et al. 2007). The significance of the SP-like immunoresponse in the bovine CL could also be explained by an interaction with the TLR-system (Fig. 3.6). Finally, the Wnt protein family, termed as a combination of Wg (wingless) and Int (oncogene int-1), is known for its numerous roles in embryonic segmentation and patterning. The Wnt protein family might share downstream interactions with the TLR family. The reason is that positive and negative feedback loops are found between TLR and Wnt, signaling affecting the expression level of NF-κb in *Drosophila* development or in macrophages exposed to mycobacteria (Blumenthal et al. 2006; Gordon et al. 2005). This feedback might be influential in the ovary, because Wnt4 knock-out mice have disorders in follicle development and in fertility (Boyer et al. 2010). It is remarkable that the systems of TLR, of complement, of tachykinin and of Wnt, belong to evolutionarily ancient families, which have changed little for 600–700 million years (Endo et al. 2006). Their development is closely associated with the appearance and function of INIM, enabling worms and jawless vertebrates to survive without adaptive immunity, which is a young branch of the immune system (Turvey and Broide 2010).

The TLR system is expressed in immune cells, and in non-immune cells such as adipocytes, mesenchymal stem cells, and also in ovarian cells of different species (Liu et al. 2008). The TLR4 on the cell surface is physically associated with CD14 and MD-2 (the molecule for LPS responsiveness of TLR4) that is a requisite for LPS signaling. Transcripts of the CD14, of MD-2 and of TLR4 are present in bovine granulosa cells (Herath et al. 2007). Another group has reported on proteins of TLR2 and TLR4 as well as on TLR8 and TLR9 in mouse granulosa cells and in cumulus cells (Shimada et al. 2006), thus lipoproteins, LPS and nucleic acids can be processed in the case of cell death. The occurrence of genes encoding co-regulatory receptors and adaptor molecules such as *CD14*, *Myd88*, *AP-1 IRF* point to the MYD88-dependent signaling cascade leading to the transcription factors AP-1, and NF-κb and the TRIF-related pathway with IRF-regulated gene activation (see above). The TLRs appear functionally active in mouse granulosa cells, because receptor activation results in specific target genes and because specific antibodies block TLR2- and TLR4-dependent chemokine secretion in cumulus cells (Shimada et al. 2006, 2008). In human granulosa cell cultures, roughly 40% of CK$^+$ cells die in a non-apoptotic manner under TLR4 activation with oxLDL (Serke et al. 2009, 2010; see Sect. 5.1). Other endogenous TLR ligands than oxLDL are fragments of hyaluronan generated from the hyaluron rich matrix during cumulus expansion (Shimada et al. 2008). Cumulus cells are also rich in HMGB1 being released from dying cells and probably acting as TLR4 ligand on DCs as reported for ischaemia-reperfusion injury of the liver (Tsung et al. 2007). Altogether, the TLR system is becoming a hot topic in research of the ovary (Richards et al. 2008, Liu et al. 2008), as the role of CK$^+$ cells comes into focus.

4.2 Localization of CK$^+$ Cells in the Intact Ovary

Granulosa cells of antral follicles are a heterogeneous population (Spanel-Borowski and Ricken 1997). The inner oocyte-associated layer differs in receptor expression and function from the mural granulosa cell layer as is nicely deduced from preovulatory follicles. Only the cumulus cells expand under the support of hyaluronic acid stabilization (Shimada et al. 2008). In follicle harvests from IVF patients, granulosa cells with angiogenic potential are characterized (Antczak and van Blerkom 2000). Human granulosa cell cultures contain a subpopulation of CK^+ cells, which lose CK filaments in the presence of LH and decrease progesterone synthesis (Ben-Ze'ev and Amsterdam 1989). The authors have related the significance of CK filaments in granulosa cells to steroidogenesis.

4.2.1
Follicles in Fetal and Adult Ovaries

According to unanimous observations, CK^+ granulosa cells exist in primordial follicles of fetal ovaries (Czernobilsky et al. 1985; Santini et al. 1993), whereas, in neonatal and adult ovaries, preantral and antral follicles lack these cells (Khan-Dawood et al. 1996; Pan and Auersperg 1998; van den Hurk et al. 1995). The origin of CK^+ follicle cells is assigned to CK^+ sex cord cells, which are populated by germ cells, and to CK^+ rete tubules after fusion with sex cords/primordial follicles (Rajah et al. 1992). Fusion and simple contact of rete tubules could coexist as two parallel events and change in frequency during ontogenesis of primordial follicles. We clearly showed the transient disappearance of CK^+ follicle cells in the adult ovary (Löffler et al. 2000). In fetal ovaries from human and cow, CK^+ sex cord cells extended from the CK^+ surface epithelium towards the medulla (Fig. 4.1a–f). Germ cells were tightly surrounded by CK^+ cells. Primordial follicles with CK^+ flat follicle cells were released from the distal end of sex cords and could contact CK^+ epithelial cells of rete tubules. In adult ovaries, the CK expression disappeared in growing preantral follicles, and sparsely reappeared in the basal region of large antral follicles. The basal localization of CK^+ cells was intensified in preovulatory follicles, where CK^+ cells were distributed throughout the mural granulosa cell layer (Fig. 4.2a–d). Regressing preantral follicles were able to resume CK expression preferentially in the inner granulosa cell layer close to the degenerating oocyte (Fig. 4.2e). Regressing antral follicles, which were characterized by an irregular granulosa cell layer and apoptotic cells, developed CK^+ cells of epithelioid or fibroblast-like appearance in the inner and outer/mural granulosa cells (Fig. 4.2f, g). The reappearance of CK^+ cells in preovulatory follicles and in follicular atresia is here explained by upcoming danger signals. In preovulatory follicles, the LH-triggered steroidogenic machinery is highly productive and coupled to the high production of ROS as by-products. They are generated by free electron leakage from the mitochondrial p450 cytochrome system (Hanukoglu 2006). The ensuing oxidative stress makes lysosomal membrane leaky for the release of caspase-3 causing cell death of granulosa cells. Indeed, roughly 20% dead granulosa cells were counted in follicle harvests from IVF women of younger age (Vilser et al. 2010). Up to 50% dead cells were found in the follicular fluid of women under high gonadotrophic treatment and with obesity. Dying and dead cells add more danger signals (S100s, HMGB1, heat shock

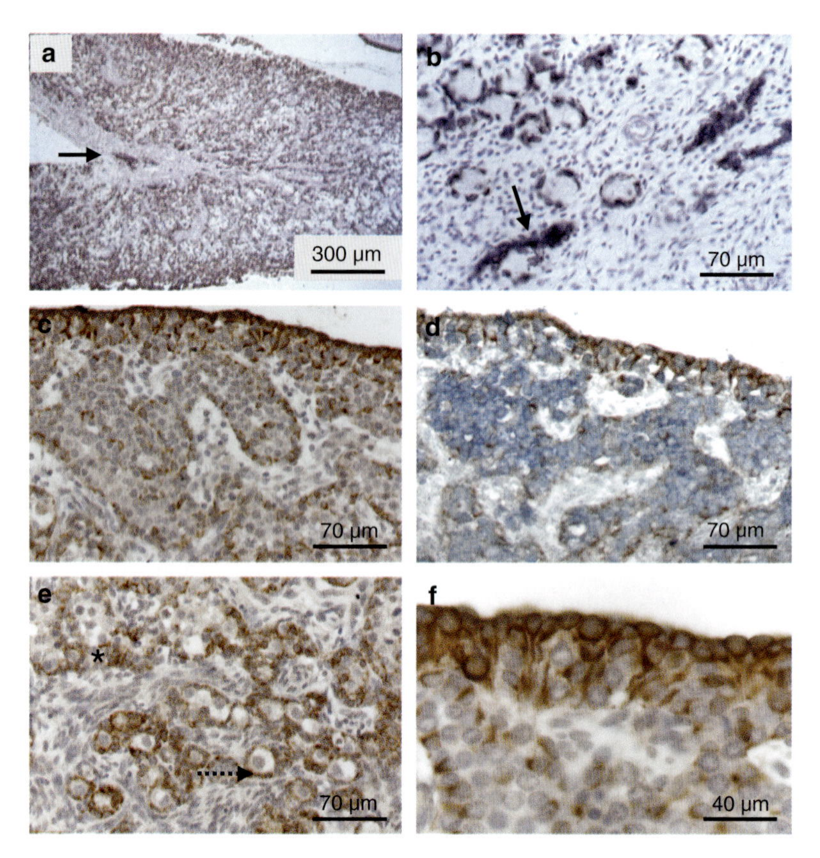

Fig. 4.1 The localization of CK^+ cells in fetal ovaries from humans (**a** and **b**) and cows (**c**–**f**) by indirect immunohistology with the mouse pancytokeratin antibody lu5. Human ovaries were allotted to the 15th–21st week of pregnancy, bovine ovaries were obtained from a fetus with a crown–rump length of 28 cm. (**a** and **b**) The cortex is well developed and displays brown CK^+ cells of the surface epithelium and of sex cords populated by germ cells. The medulla with CK^+ rete tubules (*arrow*) is small (**a**). The development from oogonia and meiotic oocytes to diplotene oocytes of primordial follicles shows a maturation gradient from the outer towards the inner zone; a primordial follicle contacts a rete tubule (*arrow* in **b**). (**c**–**f**) The CK^+ sex cord cells of brown color are continuous to the surface epithelium and sparse in comparison with prominent germ cells (**c**), which are in *blue* after double-staining the anti-Müllerian hormone with the polyclonal anti-serum from Prof. J.Y. Picard, Paris, and against CK (**d**). The outer zone confines to the surface epithelium, the middle zone to sex cords. (**e**) The inner zone shows the transition from sex cords with meiotic oocytes (*asterisk*) to primordial follicles with diplotene oocyte associated with CK^+ cells (*broken arrow*). (**f**) The outer zone consists of proliferating CK^+ surface epithelial cells enclosing oogonia negative for CK. Adapted from Löffler et al. (2000) and Tsikolia et al. (2009)

proteins; Bianchi 2007; Rock et al. 2010) to the follicle, thus increasing the danger signal pool. In regressing antral follicles, danger signals could be generated because of an insufficient microvascular bed and the inferior supply of nutrients and

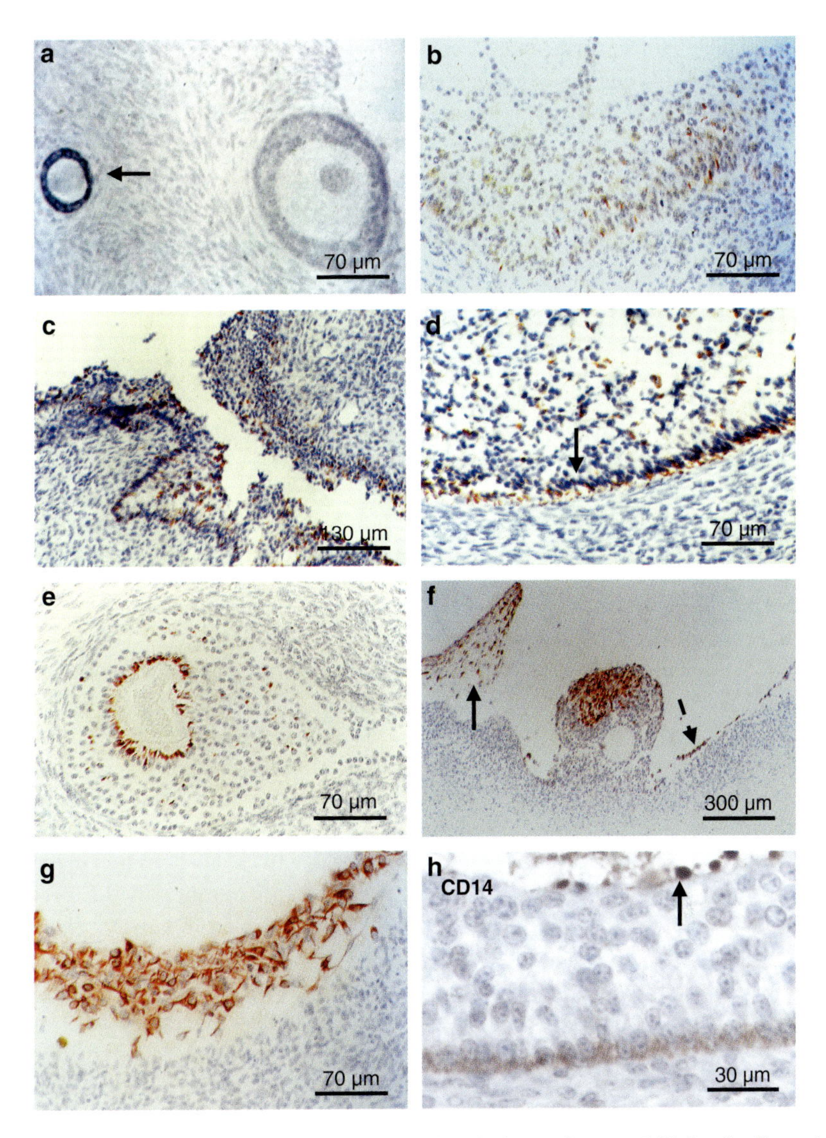

Fig. 4.2 Transient disappearance of CK expression in intact human follicles (**a**–**d**), and in regressing bovine follicles as shown with indirect immunohistology using a mouse pancytokeratin antibody lu5 (**e**–**h**). (**a**) The primary follicle on the left shows CK$^+$ follicle cells (*arrow*), not the small preantral follicle on the right. (**b**) The large antral follicle develops CK$^+$ cells in the outer granulosa cell layer. (**c** and **d**) CK$^+$ cells are noted in the gyrated granulosa cell layer of a preovulatory follicle, in particular in the basal layer (*arrow* in **d**). (**e**–**g**) The regressing preantral follicle depicts a deformed oocyte surrounded by CK$^+$ granulosa cells (**e**). (**f**–**g**) The regressing antral follicle displays CK$^+$ cells in the inner layer and in remnants of the outer granulosa cell layer (**f**). The epithelioid-type (*broken arrow*) and the fibroblast-like type (*arrow*) are labeled. An area of degenerating mural granulosa consists of fibroblast-like CK$^+$ follicle cells (**g**). (**h**) The mural granulosa cells of a regressing antral follicle reveals some apoptotic bodies (*arrow*) and CD14-positive cells in the basal layer similar to the preovulatory follicle in (**d**). Adapted from Löffler et al. (2000)

hormones. The consequences are shown as inflammatory responses (Fig. 3.2g, h). It is conceivable that danger signals are sensed through TLR in dormant CK^+ cells. In the early fetal period, a molecular cross-talk between germ cells and CK^+ epithelial cells from sex cords might have prepared gene expression of germline encoded PRR such as TLR, which is translated to the protein level in CK^+ granulosa cells in the adult period. Our search by immunostaining for TLR4-positive granulosa cells in intact preovulatory follicles has not been successful so far, yet a group of granulosa cells with CD14 expression is observed in the basal layer (Fig. 4.2f). The finding is important, because the CD14 molecule acts as co-receptor to trigger the TLR4-dependent signaling pathway (He et al. 2010; Reed-Geaghan et al. 2009). The successful co-localization of TLR4 and CD14 in CK^+ granulosa cells in serially sectioned follicles could become one small item on the long way to fully comprehend the fundamental significance of CK^+ granulosa cells as immunocompetent cells.

4.2.2 Corpus Luteum

The presence of CK^+ luteal cells is disputed for the human ovary (Czernobilsky et al. 1985; Santini et al. 1993). Summarizing all estrous cycle stages, we found no cells with CK expression in 19 out of 45 bovine CL (Ricken et al. 1995). The analysis of CL with CK expression revealed many CK^+ cells in the previous granulosa cell layer at the early developmental stage giving rise to a "zonation pattern" in the gyrated CL periphery (Fig. 4.3a, b). The cells often displayed the transition into fibroblast-like cells. Because the epithelioid and the fibroblast-like cells produced 3β-hydroxysteroid dehydrogenase (3β-HSD) activity, they were steroidogenic cells. While the CK^+ cells were high in number in the previous granulosa cell layer (61 ± 13 per 250 μm), they were small and hard to discern in the septum, being derived from the infolded thecal cell layer (Fig. 4.3c). The subsequent stage of secretion revealed uniformly distributed CK^+ luteal cells with a decreasing intensity of CK expression (Fig. 4.3d). The stage of secretion was associated with a heavy decline in CK^+ cell number (13 ± 7 per 250 μm). Stages of regression with absence of progesterone secretion were characterized by a further decrease (1 ± 1 per 250 μm) of CK^+ cells (Fig. 4.3e). One could differentiate small luteal cells (20 μm in diameter) and large luteal cells (40 μm in diameter) with CK expression (Table 4.1). The small cells dominated the developmental stage, whereas the large cells were frequent in the subsequent stages. Only the large luteal cells contained neurophysin, which was also produced by large luteal cells without CK presence (Ricken and Spanel-Borowski 1996). In view of the debate whether CK^+ cells exist in the CL (Czernobilsky et al. 1985; Santini et al. 1993), the question presently arises of a possible switch off in CK8 gene expression (Knapp and Franke 1989). The CK^+ cells in the early CL are likely derived from the CK^+ follicle granulosa cells. This is explained by their origin from CK^+ cells in sex cords and their presence in primordial/primary follicles (Figs. 4.1 and 4.2; Löffler et al. 2000). Because CK^+ cells are abundant in preovulatory follicles as well as in

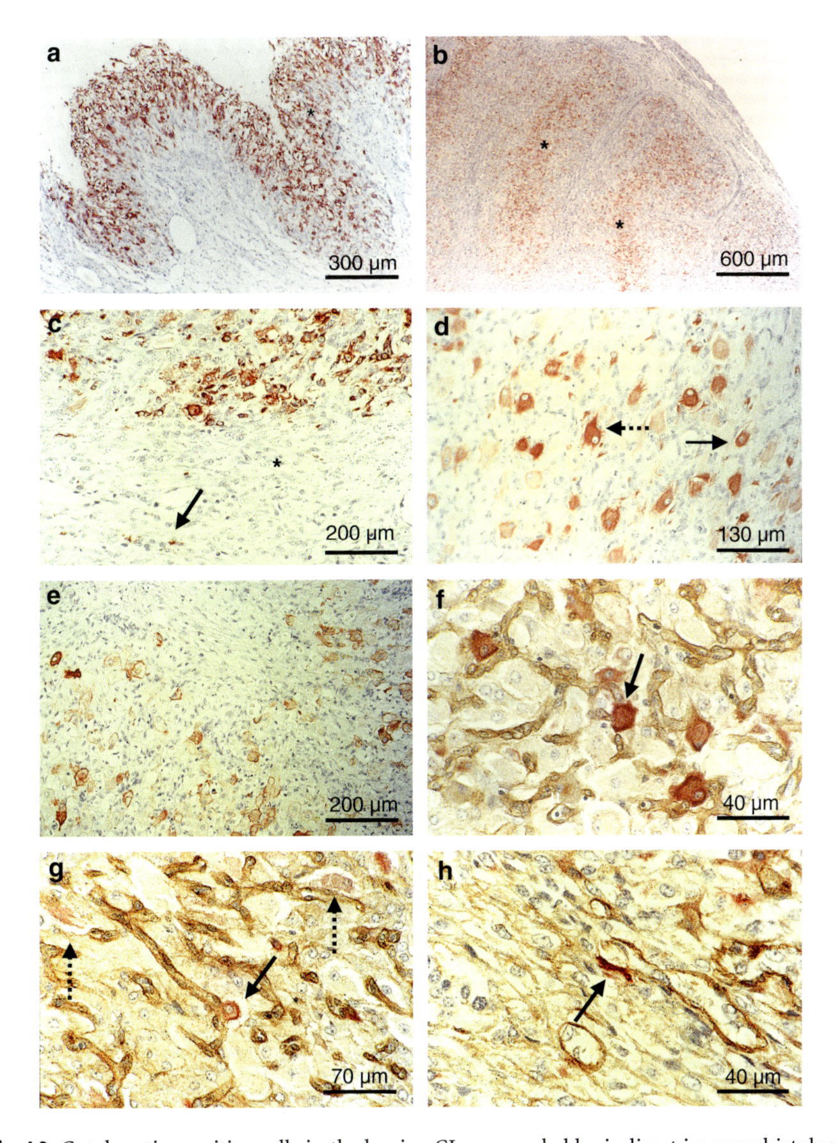

Fig. 4.3 Cytokeratin-positive cells in the bovine CL as revealed by indirect immunohistology. (**a**) CK$^+$ cells are selectively expressed in the gyrated granulosa cell layer of a freshly ruptured follicle. The fibroblast-like appearance of CK$^+$ cells is difficult to see at this magnification (*asterisk*). (**b** and **c**) The CL of early development shows "zonation" because of CK expression in the periphery, which relates to infoldings of the former granulosa cell layer (*asterisks*). A CK$^+$ cell is seldom seen (*arrow*) in the previous thecal cell layer (*asterisk*), which is forming the septum. (**d**) In the stage of secretion, steroidogenic CK$^+$ cells are small in size (*arrow*) or large (*broken arrow*); both are uniformly spread in the CL. The intensities of CK expression vary between the steroidogenic CK$^+$ cells. (**e**) In the stage of regression, the CK expression is weak. (**f**–**h**) The microvascular CK$^+$ cell is revealed by double staining for CK and for laminin in a CL of the secretory stage: epithelioid form in capillaries (*arrow* in **f** and **g**) or spindle-shaped form in a septal vein (*arrow* in **h**). Weakly stained CK$^+$ cells are noted in (**g**, *broken arrow*). Adapted from Ricken et al. (1995)

Table 4.1 Cytokeratin and/or neurophysin-positive luteal cells (LCs)

Type of expression	Early development ($n = 6$)	Stage of secretion ($n = 6$)	
	Small and large LC	Small and large LC	Large LC
Cytokeratin (CK)	74 ± 6	25 ± 6	10 ± 4
Neurophysin (NP)	63 ± 27	25 ± 7	23 ± 7
CK and NP	31 ± 9	9 ± 4	8 ± 4

Sections of bovine corpora lutea were treated with single or double immunostaining. Absolute cell numbers per area of 250 × 250 μm^2 are represented by the mean and SD. Adapted from Ricken and Spanel-Borowski (1996)

some regressing antral follicles, the LH surge cannot be the exclusive inductor. Danger sensing of oxidative stress and of cell death could trigger the renaissance of CK^+ cells by activation of a quiescent predecessor with TLR4 positivity. The suggestion is supported by a novel aspect of TLR4 biology obtained with endothelial precursor cells, which increased in number under LPS-dependent stimulation of TLR4 (He et al. 2010). An alternative way could be that oxidative stress has reactivated CK gene expression in CK^- granulosa cells.

A microvascular type of CK^+ cells can also be detected using double immunolocalization for CK and basement membrane laminin (Ricken et al. 1995). At this stage of development, isolated CK^+ cells were unspectacular in the former thecal layer (Fig. 4.3c). At the stage of secretion, the microvascular type was closely associated with capillaries and venules (Fig. 4.3f–h) being roughly 1% of the CK^+ cell pool. Because the CK^+ cells have been noted by other in microvessels of different human organs (Jahn et al. 1987; Patton et al. 1990), their ubiquitous occurrence points to a general function of the of CK^+ vascular cell type. The origin of the microvascular CK^+ cell type might be traced back to the aorta-gonado-mesonephros region in the early fetal period (Dieterlen-Lièvre et al. 2006; Pouget et al. 2006; Zovein et al. 2008; see Sect. 5.2.2 and Chap. 6).

Chapter 5
Characterization of Isolated CK$^+$ Cells

By giving enough attention to our findings on CK$^+$ cells, they remind us of hidden jewellery specialized as non lymphoid DCs to sense and to react to stressed and dying cells in the ovary. If so, CK$^+$ cells are the key player in ovarian INIM. Our exciting suggestion is supported by our own functional data on CK$^+$ cell cultures.

5.1
CK$^+$ Cells from Preovulatory Follicles with TLR4 Expression

Others have established CK$^+$ cell cultures from follicle harvests without the presence of LH. The cells have a flat and extended morphology (Ben-Ze'ev and Amsterdam 1989). We also obtained the fibroblast-like phenotype in addition to an epithelioid one, which looked like keratinocytes at the phase microscopical level and expressed CK filaments (Fig. 5.1a–d). Because both phenotypes were obtained from fresh harvests of preovulatory follicles (Löffler et al. 2000), they likely belonged to parts of the mural layer, being shed into the follicle antrum. The epithelioid type grew as a monolayer and was well demarcated by contact-inhibition from the multilayer-forming fibroblast-like granulosa cells. The demarcation of the CK$^+$ type was the prerequisite to purify the cells by mechanical selection. The CK$^+$ granulosa cells were steroidogenic as validated for CK8 and the steroidogenic acute regulatory protein (StAR, processes cholesterol within mitochondria for steroid biosynthesis) in western blots (Fig. 5.1i). The occurrence of CK$^+$ granulosa cells was accidental in culture, and could not be predicted from the follicle harvests. It might be explained by specific clinical parameters of IVF patients such as the gonadotrophin stimulation protocol, reproductive age, and obesity. This experience shows that epithelioid and fibroblast-like cells maintain CK expression after cell passage and deep-freezing.

Because the CK$^+$ cells have come into focus as danger sensors, we looked for a physiological culture model of oxidative stress. It finally turned out to be the treatment with oxidised low-density lipoprotein (oxLDL), which is a specific ligand of the lectin-like oxLDL receptor 1 (LOX-1). LOX-1 is of high importance in vascular research. The process of atherosclerosis is associated with oxLDL-dependent LOX-1 activation leading to apoptotic cell death of endothelial cells

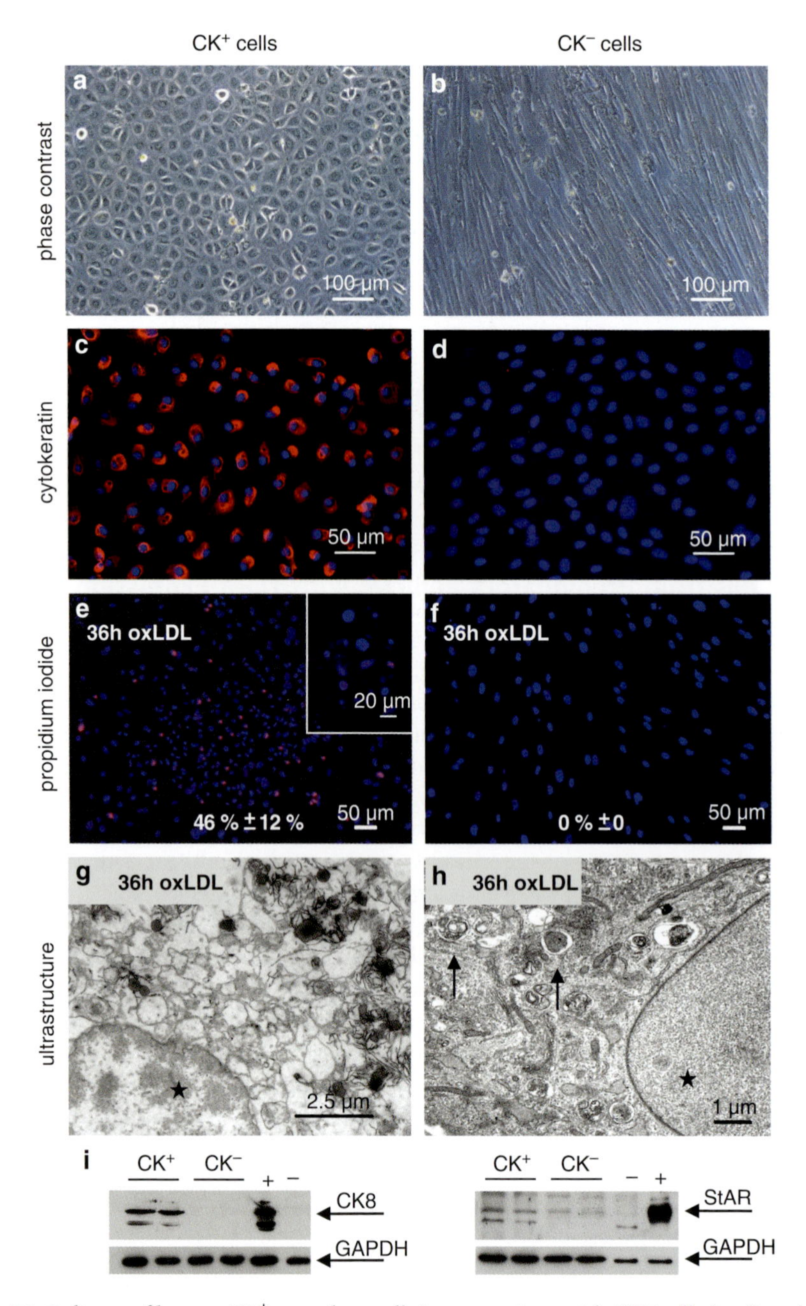

Fig. 5.1 Cultures of human CK$^+$ granulosa cells in comparison with CK$^-$ cells (**a–d**) and the reaction to serum-free treatment with 150 µg/ml oxLDL for 36 h (**e–h**). (**a–d**) Under the phase microscope, the CK$^+$ cells (*left side*) develop as an epithelioid monolayer with uniform CK expression by immunostaining with the pancytokeratin antibody lu5 (**a** and **c**). The CK$^-$ cells (*right side*) look like fibroblasts and the DAPI-stained nuclei are not associated with a

(Chen et al. 2007; Salvayre et al. 2002). The signaling pathway induces the generation of harmful ROS, which, in a vicious autocrine feedback, upregulate LOX-1 and intensify the process by more oxLDL binding (Dandapat et al. 2007; Mehta et al. 2006). Our recent discovery of LOX-1 on human granulosa cells extends its significance to the biology of the ovary. Granulosa cell cultures not classified into the CK$^+$ and CK-negative (CK$^-$) cell types underwent cell death-related autophagy. The finding correlated with the absence of apoptotic bodies and with the decrease in the protein microtubule light chain 3-I (LC3-I) indicating autophagosome formation (Duerrschmidt et al. 2006). Having purified freshly harvested granulosa cells from contaminating leucocytes, the granulosa cells expressed significantly more LOX-1 protein in younger and obese women under low gonadotrophic simulation protocol than the normal-weight counterparts (Vilser et al. 2010). Younger women also showed granulosa cells with autophagosomes both at the ultrastructural level and in western blots, which depicted an increased ratio of membrane-bound LC3-II versus cytosolic LC3-I autophagosome proteins. The augmented signs of reparative autophagy correlated with roughly 20% of dead granulosa cells in unpurified follicle harvests, whereas up to 50% were counted in the older obese group. The results encouraged us to associate the oxLDL-dependent LOX-1 activation with danger/ROS signaling. The concept has been put forward that preovulatory follicles experience extreme oxidative stress and high ROS levels (Agarwal et al. 2003, 2005). The action of ROS, which oxidizes nLDL to oxLDL, becomes likely for the preovulatory follicle. The reason is that oxLDL was higher in the follicular fluid of obese women with and without the polycystic ovary syndrome (PCOS) and thus explains ovarian disorders and infertility in obese women (Bausenwein et al. 2010). We have speculated that at this time point CK$^+$ and CK$^-$ granulosa cell subtypes are differently sensitive to an oxLDL-related LOX-1 activation. Furthermore, because ROS has a relationship to INIM function and might modulate the TLR system (Kohchi et al. 2009), we assumed that oxLDL also regulates TLR4 in the CK$^+$ and CK$^-$ granulosa cell subtypes. Oxidized lipoprotein is known as a ligand of TLR4 in macrophage and vascular function (Michelsen et al. 2004; Miller et al. 2003a).

This speculation proved to be correct. The CK$^+$ cells responded to oxLDL application by cell death (Serke et al. 2009, 2010). The percentage of 46% $\pm$ 12 of dead cells (Fig. 5.1e, f) remained unchanged under LOX-1 blockade. A comparable treatment with nLDL or with serum-deprived medium alone had no effect on the

Fig. 5.1 (continued) CK response (**b** and **d**). (**e** and **f**) The CK$^+$ cells undergo cell death without nuclear fragmentation as revealed by uptake of propidium iodide (nuclei in *pink* because of overlay with DAPI), whereas the CK$^-$ cells show no uptake; the percentage of non-apoptotic cell death is indicated (**e** and **f**). (**g** and **h**) The ultrastructure of oxLDL-treated CK$^+$ cells depicts vacuolization of the cytoplasm and chromatolysis (*asterisk*); the CK$^-$ cell shows some autophagosomes (*arrows*). (**i**) In western blot analysis, only the CK$^+$ cells express CK8; both types have the steroidogenic enzyme StAR; $\pm$ cell line MCF/and 3T3-L1 as positive/negative controls. Adapted from Serke et al. (2009)

integrity of CK^{+} cell cultures. Cell death was associated with vacuolar degeneration of the cytoplasm and chromatolysis in ultrathin sections. Cell blebbing and nuclear condensation/fragmentation as apoptotic signs were absent (Fig. 5.1g, h). In western blots, two apoptotic markers (cleaved caspase-3 and apoptosis-inducing factor) were missing. In contrast, the CK^{-} cells regulated LOX-1 under serum-free oxLDL treatment, whereas the CK^{+} cells developed only basal levels (Fig. 5.2a, b). The 36-h treatment with oxLDL (150 μg oxLDL/ml) led to survival autophagy in CK^{-} cells (Fig. 5.2c). They contained autophagosomes with the characteristic double membrane around a degenerating organelle at the ultrastructural level and a shift from the cytosolic LC3-I (18 kDa) towards the membrane-bound (16 kDa) in western blots. The oxLDL-induced autophagic response was inhibited by receptor ligation with a specific antibody. Taken together, oxLDL is associated with survival autophagy in CK^{-} granulosa cells and non-apoptotic death of CK^{+} granulosa cells. Survival autophagy in CK^{-} cells depends on LOX-1 regulation.

Next, the TLR4 system was upregulated in the CK^{+} cells when the cultures were treated with oxLDL under serum-free and endotoxin-free conditions. The TLR4 increase was accompanied by the strong CD14 immunoresponse (Fig. 5.3a–d). These point to the functionally active TLR4, which requires heterodimerization of CD14 as co-regulatory molecule (He et al. 2010; Miller et al. 2003b). The CD14 upregulation in the oxLDL-treated CK^{+} cell cultures also underlines the significance of the finding that CD14-positive granulosa cells are localized in the antral follicle (Fig. 4.2h). They might represent CK^{+} cells with the expression of TLR4 and signal receptor upregulation because of an environment under oxidative stress. When a neutralizing antibody-blockade of TLR4 was conducted in oxLDL-treated cultures, the percentage of dead CK^{+} cells increased strikingly (Serke et al. 2009). Changing combinations of TLR4 and co-regulatory receptors could explain the increased cell death response (Seimon et al. 2006). Low ROS levels (determined with MitoSOX™ Red from Molecular Probes; the reagent is oxidized by mitochondrial superoxides into a red fluorochrome measured by flow cytometry; Serke et al. 2009) in oxLDL-treated CK^{-} cells correlated with high StAR production, whereas the reverse was stated for CK^{+} cells (Fig. 5.4a–c). Thus, the CK^{+} cells are more sensitive to danger/oxLDL signals than CK^{-} cells. The CK^{+} cells can be considered as danger sensors and might contribute widely to the 20–50% of dead granulosa cells in fresh follicle aspirates (Vilser et al. 2010).

Cumulus cells provide another source of CK^{+} cells. When they were obtained from cumulus oophorus complexes by hyaluronidase digestion, the fibroblast-like CK^{+} phenotype regularly formed a network hidden among CK^{-} cells (Fig. 5.5a, Serke et al. 2010). The epithelioid type was not seen. The hidden network of fibroblast-like CK^{+} cells did not allow the mechanical selection of CK^{+} cumulus cells. The cumulus cell cultures, thus a mixed population, revealed no regulation of LOX-1 and of TLR4, but the increased expression of CD36 under serum-free oxLDL treatment for 36 h. The CD36 increase was associated with 7.9% ± 1.8% of dead cells. It has been recently reported that the oxLDL binding to CD36 precedes heterodimerization of TLR4 and TLR6 in sterile inflammation of atherosclerosis

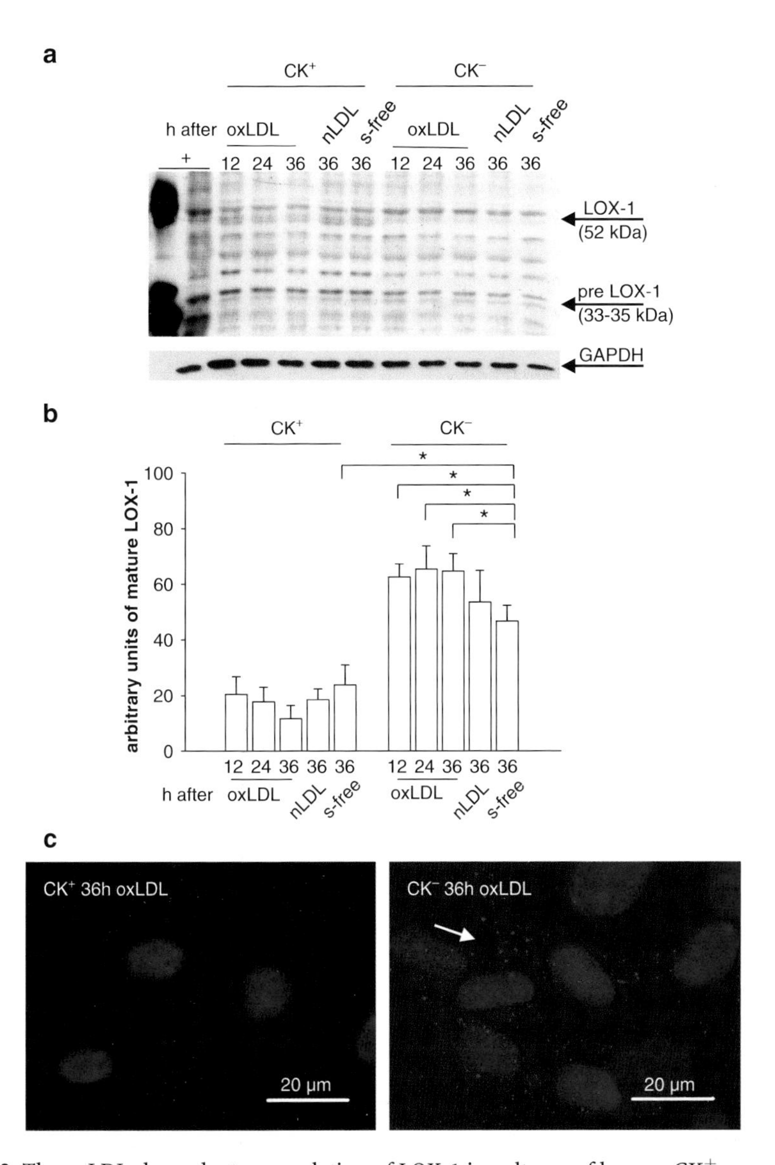

Fig. 5.2 The oxLDL-dependent upregulation of LOX-1 in cultures of human CK^{+} granulosa cells in comparison with CK^{-} cells. Cultures were treated with 150 μg/ml oxLDL/nLDL under serum-free (*s-free*) conditions; + = positive control with recombinant LOX-1. (**a**) The CK^{+} cells produce less LOX-1 than the CK^{-} cells in the representative western blot. (**b**) The CK^{+} cells lack an upregulation of LOX-1 in the semiquantitative evaluation of three independent experiments; values represent mean ± SEM of three independent experiments; $*p < 0.05$. (**c**) The immunostaining for the light chain 3 protein (LC3, a marker for autophagosomes) reveals many autophagosomes in the CK^{-} cells (*arrow*). Adapted from Serke et al. (2009)

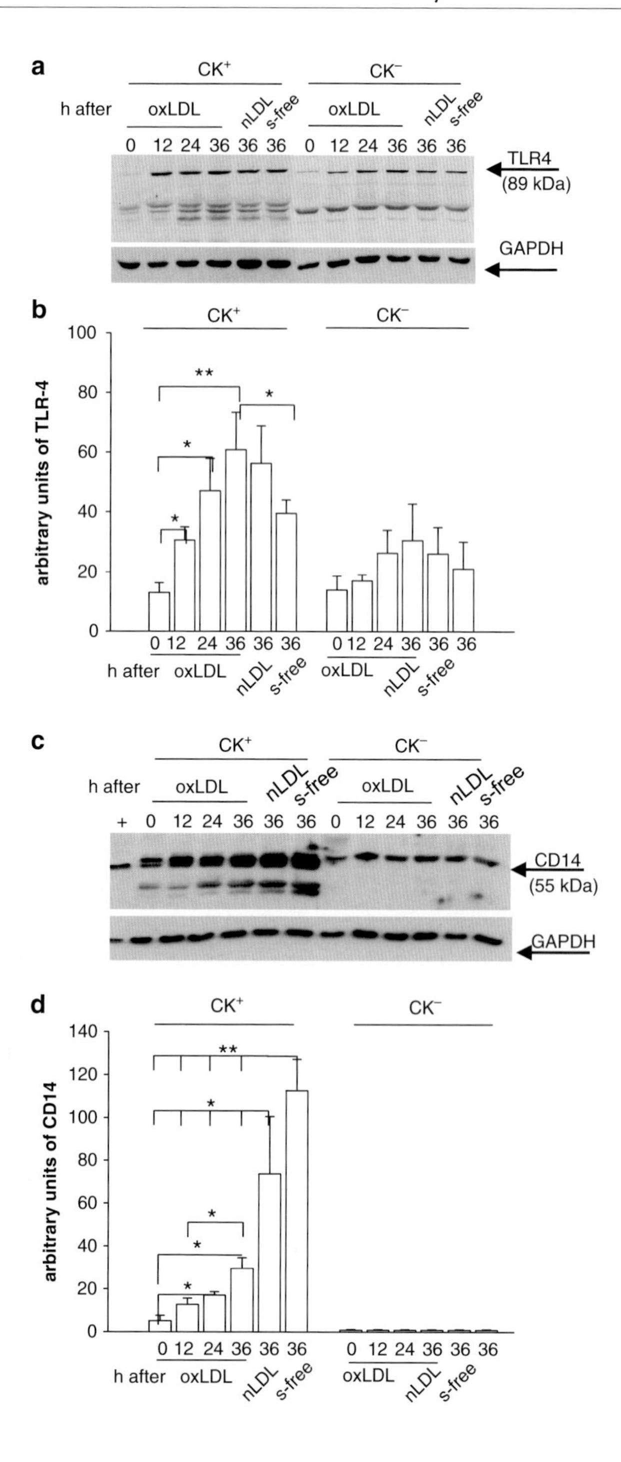

a
CK⁺
CK⁻
h after
oxLDL
nLDL
s-free
0 12 24 36 36 36 0 12 24 36 36 36
TLR4
(89 kDa)
GAPDH
b
arbitrary units of TLR-4
c
+
CD14
(55 kDa)
d
arbitrary units of CD14

(Stewart et al. 2010) and that TLR4 could be involved in the regulatory loop between cumulus cells and sperms during fertilization in the Fallopian tube (Richards et al. 2008; Shimada et al. 2008). We have also investigated the anti-oxidant levels of the three granulosa cell subtypes in response to oxLDL treatment. Indeed, a ranking in activities occurred: high levels of superoxide dismutase in CK^+ cells, of catalase in cumulus cells, and of the glutathione system in CK^- cells (Fig. 5.5b, c; Serke et al. 2010). The corresponding supernatants contained high levels of catalase activity both for CK^+ cells and for cumulus cells. This leads to the conclusion that catalase activity in the supernatant arises through leaky membranes and thus indicates cell damage caused by oxLDL-dependent oxidative stress. The catalase activity increase, which was observed in the follicular fluid of obese women with fertility disorders compared to normal-weight women (Bausenwein et al. 2010), might thus indicate augmented oxidative stress. Changes in catalase activities can thus be established as a scale of danger.

Collectively, the CK^+ granulosa cell subtypes are highly susceptible to oxLDL-treatment. They seem to sense danger signals such as oxidative stress and respond to the danger/oxLDL signals. The TLR4-dependent cell death and the catalase activity increase in the microenvironment might be one aspect among others. Additional TLRs such as the endosomal TLR7-9, which recognise nucleic acids, have to be considered. Insights into the influence on TLR activation by co-signaling through complement receptors or tachykinin receptors are required (see Sect. 4.1; Hajishengallis and Lambris 2010; O'Connor et al. 2004). A successful outcome could reinforce our own findings on the tachykinin–tachykinin receptor system in the bovine ovary (Figs. 3.5 and 3.6) and support the statement that non-neuronal tachykinins play a role in ovarian function (Debeljuk 2006). The TLR-mediated Wnt pathway is reported to be a key process for sustained macrophage activation (Pereira et al. 2009), and might also be important in the ovary. WNT4 is required for follicle growth and fertility (Boyer et al. 2010). We need to know whether CK^+ granulosa cells capture and process antigens by MHC regulation under treatment with IFN-γ and TNF-α treatment as do DCs (Banchereau and Steinman 1998; Mellman and Steinman 2001). Cytokine stimulation of the CK^+ granulosa cells might result in specific profiles of cytokines and chemokines, which are influential on growth and function of tyrosine-kinase receptor (KIT)-positive thecal cells. They are limited to the periphery of the developing bovine CL and are thus close neighbors to the CK^+ cell zone (Fig. 4.3; Ricken et al. 1995; Spanel-Borowski et al. 2007). One has to understand why CK^+ cells are missing in preantral and antral

Fig. 5.3 The oxLDL-dependent upregulation of TLR4 (**a** and **b**) and CD14 (**c** and **d**) in cultures of human CK^+ granulosa cells in comparison with CK^- cells. Cultures were maintained endotoxin-free, treated with 150 μg/ml oxLDL/nLDL under serum-free (*s-free*) conditions and western blot analysis done (representative blots in **a** and **c**, arbitrary unit measurements in **b** and **d**); + = positive control, values represent mean ± SEM of three independent experiments; $*p < 0.05$, $**p < 0.001$. Adapted from Serke et al. (2009)

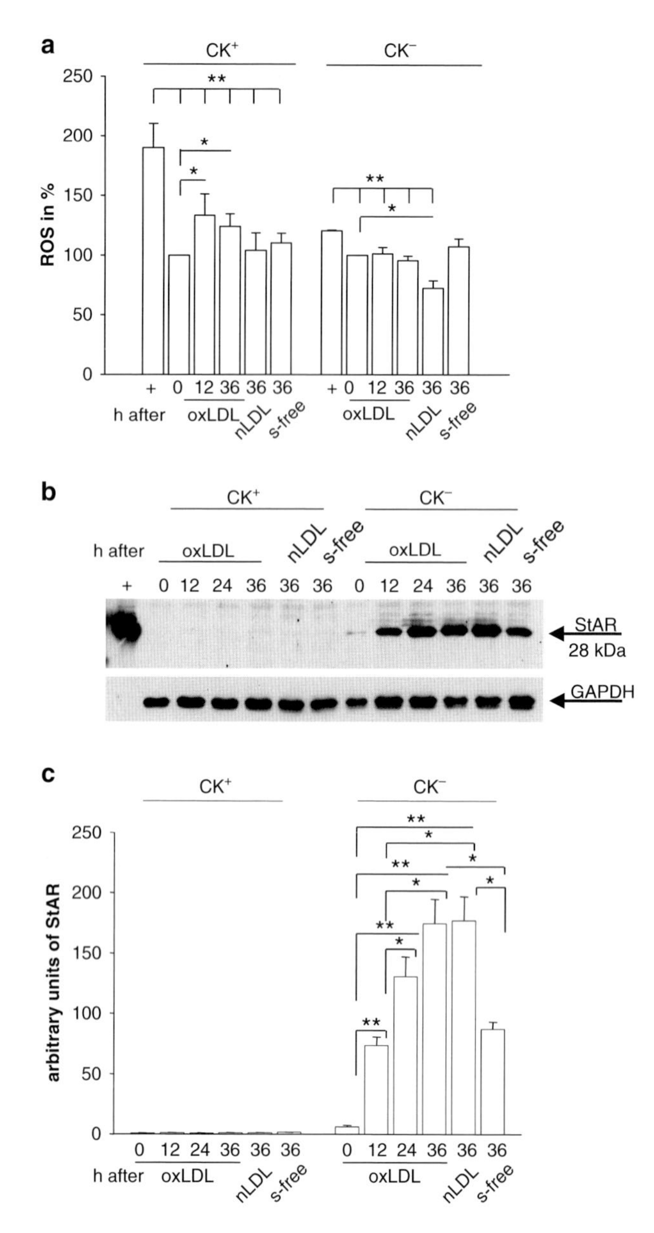

Fig. 5.4 The generation of ROS (**a**) and the sensitivity of the StAR enzyme in human CK^{+} granulosa cells in comparison with CK^{-} cells. Cultures were treated with 150 μg/ml oxLDL/nLDL under serum-free (*s-free*) conditions; + = positive control; values represent mean ± SEM of three independent experiments; $*p < 0.05$, $**p < 0.001$. (**a**) ROS relate to mitochondrial superoxides, which were detected with the MitoSOX™ Red (Molecular Probes) and measured with flow cytometry. An increase of ROS with 12 h of oxLDL

follicles. The CK$^+$ cells could be dormant and small in number until their proliferation is induced in the preovulatory follicle. It is also conceivable that the CK$^+$ cells are concealed in preantral and antral follicles because of a cytoskeletal change. This could be caused by a switch-off in *CK* genes (see Sect. 5.2.2 and Chap. 6).

5.2 CK$^+$ Cells from CL in Comparison with CK$^-$ Cells

The CL life cycle compares with a big remodeling event. The transient endocrine organ develops, becomes fully functional for some days in cows and humans, stops progesterone secretion abruptly, and then decays over several ovarian cycles. The remodeling process requires a tight control in leucocyte recruitment, in angiogenesis, and in tissue repair by phagocytosis and connective tissue substitution. In golden hamsters, a CL life completely disappears through rapid turn-over of apoptotic bodies at the end-point of a 4-day estrous cycle (Fig. 2.1). In other small rodents, apoptotic bodies are rarely seen and CL regression requires several ovarian cycles. No histological traces are retained of the former CL in golden hamsters and in small rodents. In cows and humans, the ovarian cycle extends over 3–4 weeks and, in the secretory phase, the CL seems to depend on survival autophagy (Del Canto et al. 2007; Gaytn et al. 2008). The CL becomes a corpus albicans, a hyalinized scar, which remains for a long time. Much work has been achieved on various cells and intra-ovarian factors, which control the complex process of CL transformation, maintenance and regression (Niswender et al. 2000; Stouffer et al. 2007; Schams and Berisha 2004). The molecular pattern is also revealed at the gene and protein levels in the human (Devoto et al. 2009; Stocco et al. 2007). The open question transfers to the search for the endocrine precursor cell, which multiplies and differentiates in the CL of pregnancy. Likewise, the remarkable healing process after follicle rupture is widely neglected in early CL development. The CK$^+$ cell as potential DC in the CL could coordinate the exceptional repair task.

Like granulosa cells, the cell population in a CL is heterogeneous. The majority of cells represent vascular cells from arterioles, capillaries and venules, although steroidogenic cells are more prominent in size (Niswender et al. 2000; Schams and Berisha 2004, Stouffer et al. 2007). The steroidogenic cells divide into small and large luteal cells from the former thecal and granulosa cell layers, respectively. Each isolation and cultivation procedure thus faces the necessity to characterize the potential cells correctly. This project has been successfully conducted for the

Fig. 5.4 (continued) treatment is seen in CK$^+$ cells; no changes of ROS levels occur in the CK$^-$ cells in respect to the s-free control. (**b** and **c**) The representative western blot and the evaluation of arbitrary units shows extremely low levels of StAR for CK$^+$ cells throughout the experiment, while the steroiodgenic enzyme strikingly increases in CK$^+$ cells. Adapted from Serke et al. (2009)

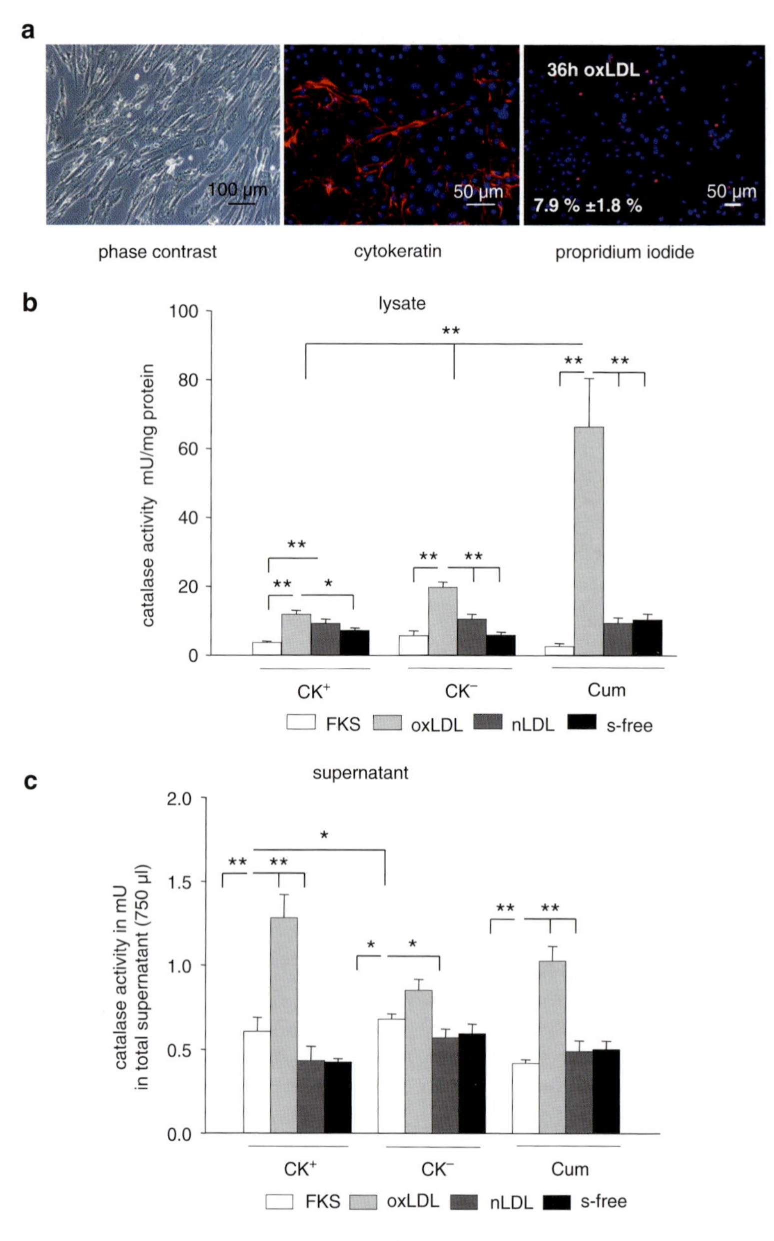

Fig. 5.5 Human cumulus cells (Cum) with CK$^+$ and CK$^-$ cells in (**a**) and catalase activity in *CK*$^+$ cells, *CK*$^-$ cells and *Cum* (**b** and **c**). Cultures were treated with 150 µg/ml oxLDL/nLDL under serum-free (*s-free*) conditions; serum-containing medium (*FKS*) was used as control; values represent mean ± SEM of at least ten independent experiments; $*p < 0.05$, $**p < 0.001$. (**a**) Cumulus cells form a spindle-shaped monolayer under the phase contrast microscope (*left*), contain a network of CK$^+$ cells in immunostained cultures, and a minor

bovine CL (Spanel-Borowski 1991; Spanel-Borowski and van der Bosch 1990; Davis et al. 2003; Fenyves et al. 1993). Five different phenotypes were primarily defined as endothelial cells. They depicted the classical endothelial cell features such as monolayers indicating contact-inhibited growth, expression of FVIIIr antigen and the uptake of Dil-acLDL. *Cell type 1* demonstrated an isomorphic cobble-stone pattern; the epithelioid appearance was reminiscent of keratinocytes and related to a well-developed network of CK filaments and to the peripheral ring of actin filaments (Fig. 5.6a–c). The diffuse perinuclear staining of FVIIIr antigen was untypical for endothelial cells, whereas the lysosomal-associated uptake of Dil-acLDL was similar to endothelial cells (Fig. 5.6d, e). Because α_2-macroglobulin (α-2M) characterizes endothelial cells from arterioles (McAuslan et al. 1982), the protein presence in CK^+ type 1 cell (Fig. 5.6f) could indicate the source of origin. α_2-macroglobulin has evolved from the same ancestral gene as C3 and C4 (Doan and Gettins 2007), and the protein is also produced by granulosa cells as is the α-2M receptor (Ireland et al. 2004). Apart from the classical role of proteinase inhibition and growth factor binding, α-2M is reported to do more. It interacts with MBL, which is then able to activate the classical complement pathway (Arnold et al. 2006). Nanomolar concentrations of α-2M decrease cell proliferation and migration of astrocytoma cells by impeding Wnt/β-catenin signaling (Lindner et al. 2010). It is thus possible that beyond protease inhibition α-2M is involved in sterile inflammation and tissue repair in periovulatory structures. Whether the ultrastructural granules with a matrix of low density in cell type 1 compared with the α-2M-positive granules in immunostained sections (Figs. 5.6f and 5.7f) remains to be clarified. It is also necessary to investigate whether C3 is generated by CK^+ cells in view of its association with the α-2M family (Köhl 2006b). *Cell type 1* was endowed with intercellular junctions such as gap junctions, tight junctions and adherens junctions (Figs. 5.6g, h and 5.7a; Mayerhofer et al. 1992; Fenyves et al. 1993; Ricken et al. 1996).

The rich array of microvilli in the scanning electron microscope alludes to a high resorptive activity of CK^+ cells, and the single/primary cilium allows speculations on its function (Fig. 5.7b–e). The single cilium is non-motile because it lacks a central microtubule pair. The single cilium is present in almost all vertebrate cell types. Here, the single cilium arose from a deeply invaginated cytoplasm

Fig. 5.5 (continued) percentage of dead cells with pink nuclei (because of overlay by DAPI stained nuclei) after uptake of red propidium iodide. (**b**) Catalase activity was determined using the Catalase Assay Kit (Cayman Chemical, Michigan, USA). The increase in catalase activity is at its maximum in oxLDL-treated cumulus cell lysates in comparison with the CK^+ and CK^- cells. (**c**) The catalase activity in supernatants, which is strong in oxLDL-treated CK^+ cells and in cumulus cells, appears to be higher in CK^+ cell cultures. Because of higher percentages of dead cells in oxLDL-treated CK^+ cells (compare with Fig. 5.1e) and lower percentages in cumulus cells (in **a**, *right*), catalase activity might come from dead cells by passive release. Adapted from Serke et al. (2010)

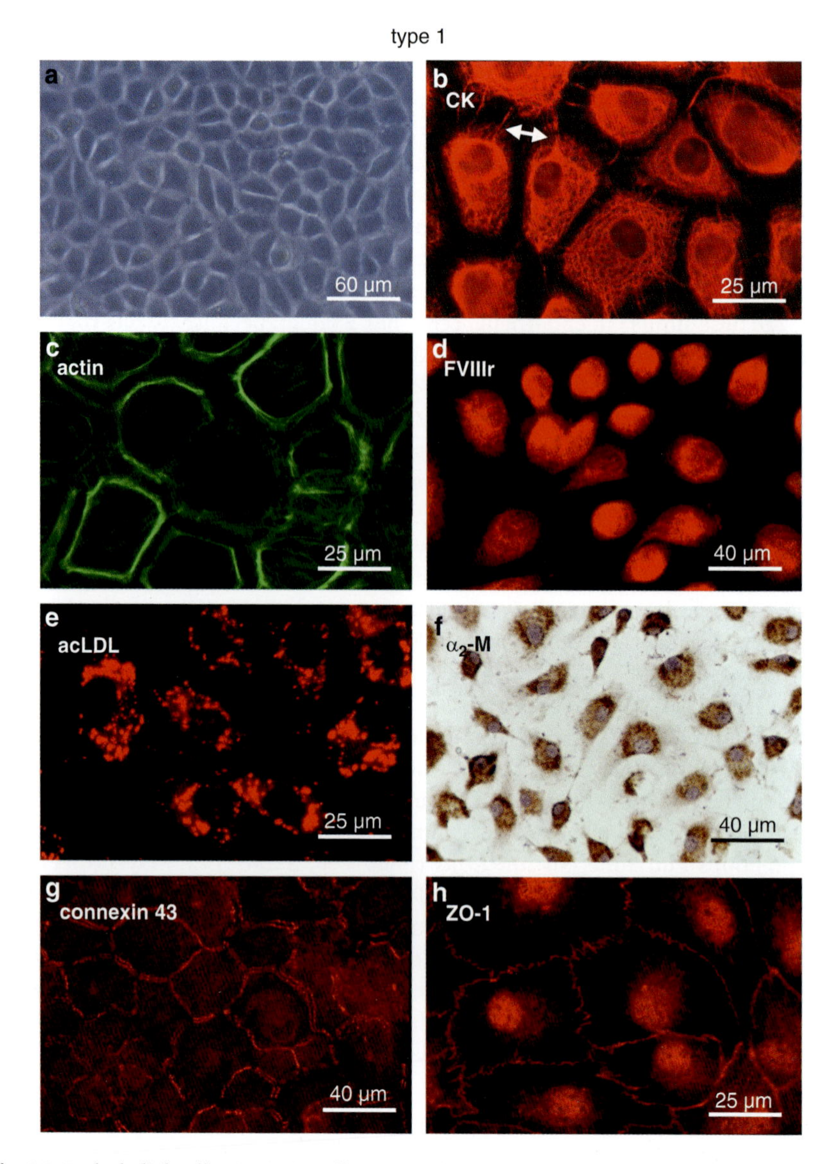

Fig. 5.6 Endothelial cell criteria in cell type 1 cultures from the bovine CL under the phase contrast microscope (**a**), by localization with immunofluorescence/immunocytology after fixation with absolute ice-cold ethanol for 5 min (**b, d, f–h**), by staining with phalloidin-FITC after 20-min fixation with 2% paraformaldehyde containing 0.2% Triton-X for permeabilization (**c**), and by uptake of Dil-acLDL (10 µg/ml in serum-containing medium at 37°C for 4 h in **e**). (**a**) The isomorphic and epithelioid monolayer of cobble-stone appearance indicates contact-inhibited growth. (**b–f**) Each cell develops a dense network of CK filaments; the peripheral bundles in contact with the neighbor is reminiscent of a desmosome (*two-sided arrow* in **b**); the CK^+ cells are characterized by a peripheral ring of

membrane (Spanel-Borowski 1991; Wolf and Spanel-Borowski 1992). The out-of-date opinion says that the single cilium represents a non-functional evolutionary relict. The modern opinion considers the single cilium as a mechano-and chemo-sensory organelle, and the ciliary proteins are implicated in signal transduction and morphogenetic pathways (Davis et al. 2006; Singla and Reiter 2006). The single cilium in the CK^+ cells could be qualified to sense danger signals by enriched TLR presence, and the increase in danger signals in the moat-like environment could preferentially provoke the TLR-dependent signaling cascade. The CK^+ cells are thus a beautiful culture model to validate the possibility, and to judge the ubiquitous occurrence of single cilia as antenna of INIM. Next, the CK^+ cells appear to have an unconventional way of protein secretion (Fig. 5.8). The immunogold technique conducted for the lectin localization of concanavalin A (ConA) in non-permeabilized cells revealed solitary ConA-positive circle-like forms of 2.5 μm diameter (Herrman et al. 1996). The forms were missing after pre-incubation with the ConA-inhibiting α-mannopyranoside at 0.2 M concentration. The ultrathin section revealed the ConA-positive forms as discontinuities of the plasma membrane together with shedding of ConA-positive vesicles. The observation is explained by ectocytosis. It is understood as the non-classical route of protein trafficking through exosomes, thus not involving the rough endoplasmic reticulum and the Golgi-apparatus (Keller et al. 2008, Zhan et al. 2009). Ectocytosis could stand for the unclear mechanism of how leaderless proteins such as IL-1α, caspase-1, and b-FGF are secreted during inflammation and tissue repair. Of note, many danger signaling molecules, i.e. the alarmins such as heat shock proteins, HMG1, IL-1α, serum amyloid and S100 proteins, lack a leader signal and are secreted via a non-classical pathway (Bianchi 2007; Rock et al. 2010). Because oxLDL causes the secretion of HSP70 in endothelial cells (Zhan et al. 2009), oxLDL might do the same to CK^+ granulosa cells and mediate the presence of alarmins in preovulatory follicles (Richards et al. 2008). In summary, the presence of CK^+ cells could correlate with S100-positive granulosa cell subtypes (unpublished) seen in sections of polycystic and post-menopausal ovaries. They were immunostained for S100-positive nerve fibers (Heider et al. 2001).

The polymorphic *cell type 2* monolayer also consisted of CK^+ cells, and additionally contained a subpopulation of desmin-positive cells (Fig. 5.9a, b), which were defined as vascular smooth muscle cells (Spanel-Borowski 1991). The desmin-positive cells seemed to be involved in the generation of three-

Fig. 5.6 (continued) actin filaments in the cytoplasm (**c**), a diffuse and central response by FVIIIr antigen-positive granules (**d**), a granular-like pattern because of Dil-acLDL in lysosomes (**e**), and by α_2M-positive granules (**f**). (**g** and **h**) The continuous punctuated line of the connexin 43 antigen indicates gap junctions (**g**). The continuous straight line of the ZO-1 antigen corresponds to tight junctions (**h**). Adapted from Spanel-Borowski and van der Bosch (1990), Mayerhofer et al. (1992), Fenyves et al. (1993) and Ricken et al. (1996)

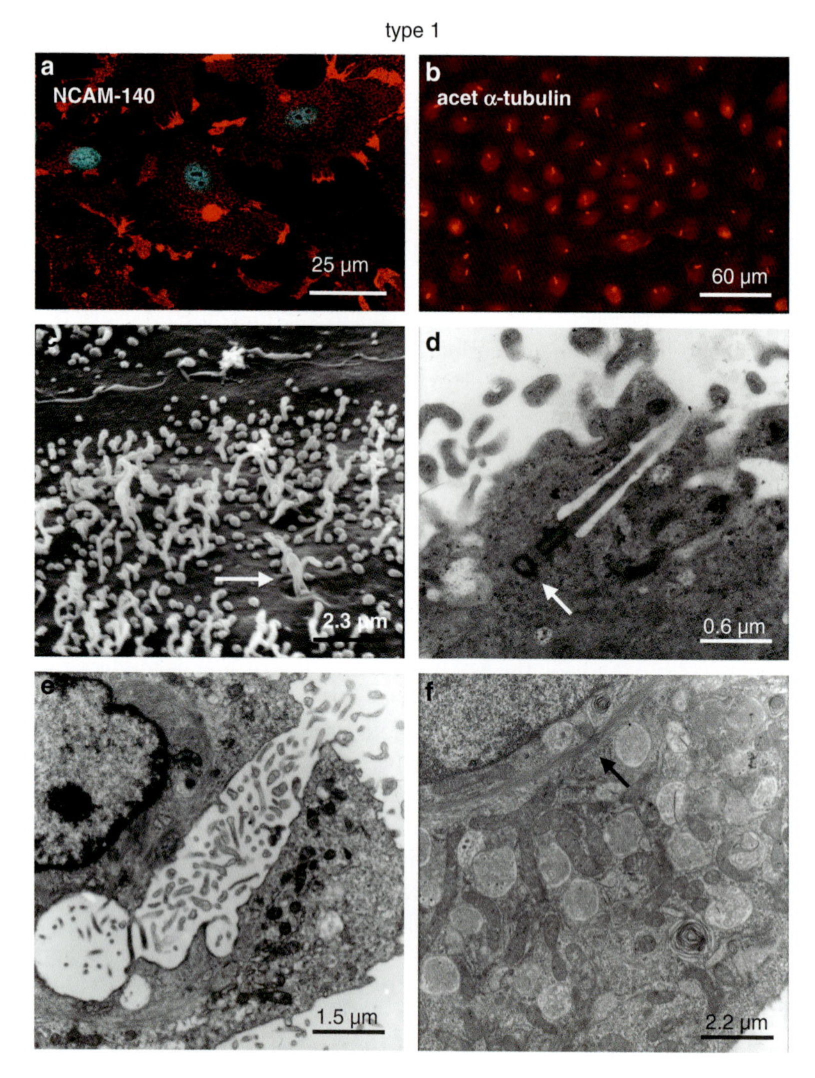

Fig. 5.7 Special features of cell type 1 cultures from the bovine CL seen by immunofluorescence localization after fixation with ice-cold absolute methanol for 5 min (**a**), with 2% paraformaldehyde and 0.25% glutaraldehyde in microtubule-stabilizing buffer before treatment with absolute methanol at −20°C (**b**), or for the study of ultrastructure after fixation with 2% glutaraldehyde for 1 h followed by post-fixation with 1% osmium tetroxide (**c–f**). (**a**) The periphery depicts NCAM-140-positive patches reminiscent of adhesion plates. (**b**) Each cell develops a single cilium, which is positive for acetylated α-tubulin and 6–10 μm in length. (**c**) The cell is rich in microvilli and shows a single cilium surrounded by a "moat" under the scanning electron microscope (*arrow*); cells were dried by the critical-point drying method and coated with palladium. (**d**) The shaft of a single cilium with the basal body and the basal foot (*arrow*) is surrounded by a deeply invaginated cell membrane. (**e**) The deep invagination of the cell membrane, which is rich in microvilli, might correspond to the lateral portion of the

dimensional tubules, which regularly developed in post-confluent type 2 cultures within 2–4 weeks. Subordinate tubules with a braided appearance under the scanning electron microscope arose from a principal branch within the monolayer. Tubules represented an inside-out model. The tubule center contained the extracellular matrix, whereas the apical side with distinct cell borders faced the culture medium (Fig. 5.9c, d). Some desmin-positive cells depicted granules of low electron density and well-developed desmosomes (Fig. 5.9e, f). They probably supply firm cell cohesion, which was incompletely disrupted by mechanical disintegration at the time of cell isolation from the bovine CL. The presence of braided-like tubules in the cell type 1 monolayer (Spanel-Borowski and van der Bosch 1990) is presently judged to be caused by the presence of desmin-positive cells.

Cell types 3 and 4 formed a monolayer with either spindle-shaped cells or round cells. The pattern was associated with prominent or delicate aster-like actin filament patterns extending throughout the cytoplasm (Fig. 5.10a–d). The FVIIIr antigen associated with the Weibel-Palade granules specific for endothelial cells, and Dil-acLDL was present in lysosomes (Fig. 5.10e, f). Microvilli were absent on the cell surface. In type 3 cultures, tubules developed spontaneously in post-confluent cultures (Fig. 5.10g). They were loosely tethered to the monolayer, and lacked cell borders. In type 4 cultures, pseudotubules formed a two-dimensional network (Fig. 5.10h). At the beginning of the phenotype classification, it was not in our mind to have obtained steroidogenic cells from the CL, because luteal cells are highly differentiated, non-proliferative and non-adherent during cultivation. It had to be learnt that *cell type 5* behaved like granulosa cells in comparing morphology and function (Spanel-Borowski et al. 1994a). Although type 5 cells grew as a flat monolayer at the phase microscope level indicating contact-inhibition, cell borders with conspicuously long filipodia overlapped at the ultrastructural level (Fig. 5.11a, e, f). The circular band of actin filaments and the presence of the neuronal adhesion molecule (NCAM-140) were reminiscent of the expression pattern in type 1 cells (compare Fig. 5.11b, c with Figs. 5.6c and 5.7a). Type 5 cells produced the cholesterol side chain cleavage enzyme P450scc also found in granulosa cells and in CL lysates and absent in types 1–4 (Fig. 5.11d, h). The intracellular accumulation of lipid droplets was another sign of steroidogenesis (Fig. 5.11g). The fact that type 5 cells represent granulosa-like cells breaks up the dogma that all follicle cells transform into luteal cells. Type 5 cells obviously have escaped the process of luteinization at the time of follicle transition into a CL. For many years, our concept has been that type 5 cells represent precursor cells, which differentiate into luteal cells in case of CL maintenance during pregnancy. Yet, experiments have failed to transform the granulosa-like cells into fully luteinized

Fig. 5.7 (continued) cilium moat. **(f)** The cell contains many granules with an amorphous matrix of low electron density and a perinuclear filament ring (*arrow*). Adapted from Spanel-Borowski (1991), Spanel-Borowski and van der Bosch (1990) and Wolf and Spanel-Borowski (1992)

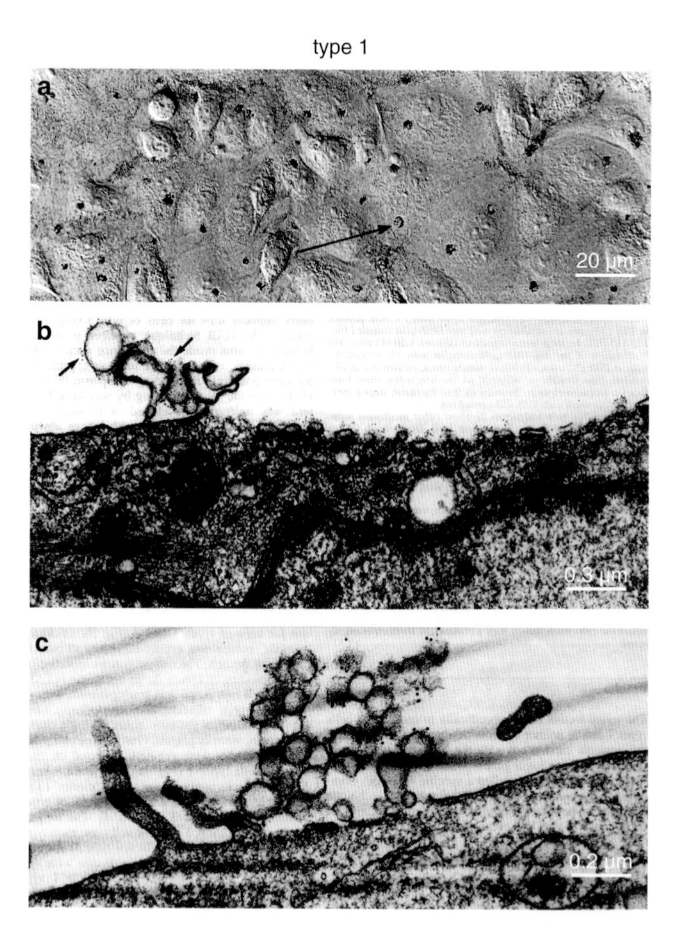

Fig. 5.8 Signs of ectocytosis in cell type 1 cultures from the bovine CL revealed by immunogold labeling of concanavalin A (ConA) binding sites after fixation with 2.5% glutaraldehyde for 30 min, lectin incubation at 4°C overnight, and incubation with streptavidin-horseradish peroxidase (HRP), followed by an anti-HRP gold complex and silver enhancement (**a**) or embedding in Epon 812 (**b** and **c**). (**a**) The non-permeabilized cells disclose single circles of ConA positivity (*arrow*). (**b**) In the area of a discontinuous plasma membrane, electron-lucent vesicles have accumulated. The outside vesicle membrane is labeled with 6-nm gold particles (*arrows*), not the intact cell membrane. (**c**) Similar to (**b**), yet with 12-nm gold particles. Adapted from Herrman et al. (1996)

luteal cells. Perhaps, type 5 cells do not express receptors for LH and FSH as shown for cell type 1 at the mRNA level (Aust et al. 1999).

As regards the yield of five phenotypes from the bovine CL of different functional stages, cell type 1 and 2 colonies were comparable in number for different CL stages (Fig. 5.12). For types 3 and 4, the stage of secretion gave a

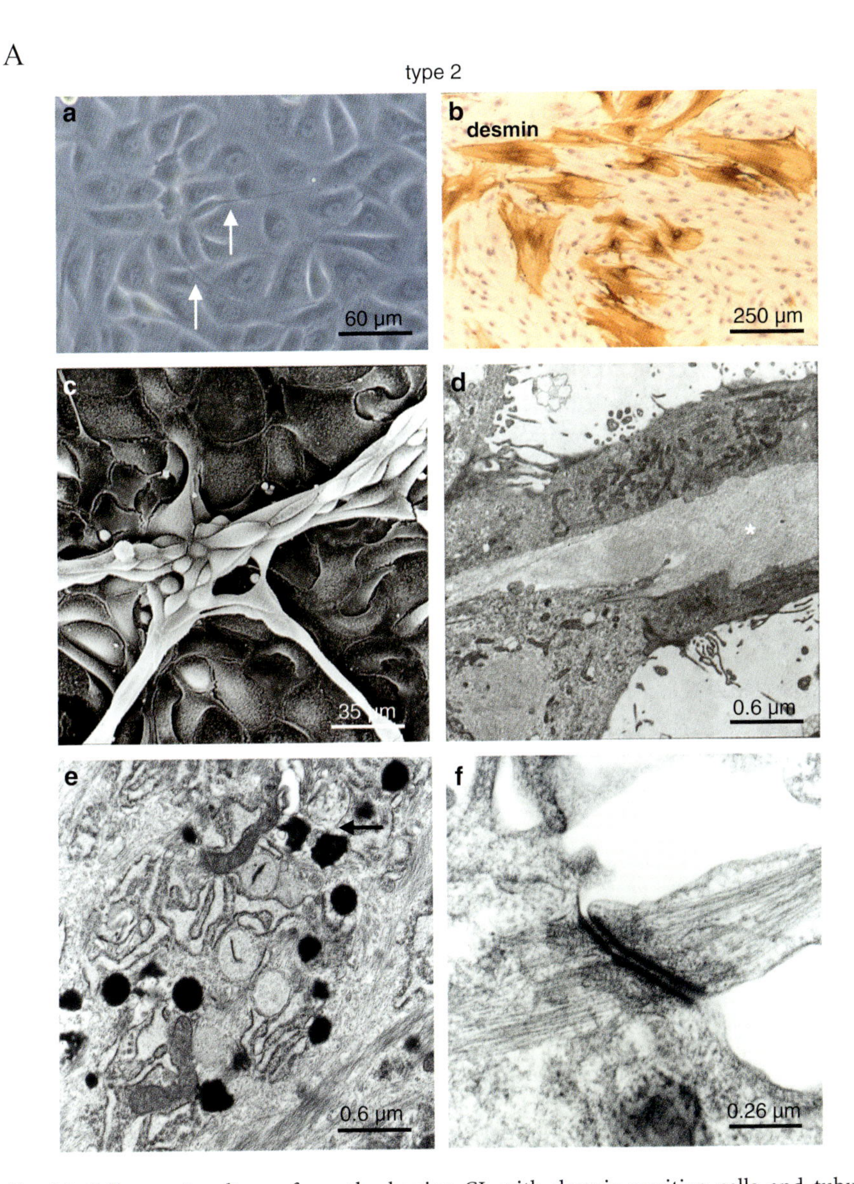

Fig. 5.9 Cell type 2 cultures from the bovine CL with desmin-positive cells and tubule formation, seen under the phase contrast microscope (**a**), by immunocytology after fixation with 2% paraformaldehyde (**b**) and for the study of ultrastructure after fixation with 2% glutaraldehyde for 1 h and post-fixation with 1% osmium tetroxide (**c–f**). (**a**) The epithelioid and polymorphic monolayer discloses cells with a long filipodium extending towards a distant cell (*arrows*). (**b**) Desmin-positive cells are revealed as vascular smooth muscle cells. (**c**) The stem of the dendroic tubule with distinct cell borders is anchored within the monolayer. (**d**) The longitudinal cut through a tubule reveals the core as extracellular matrix (*asterisk*), and microvilli at the cell surface. (**e**) The smooth-muscle-cell type with

much better output in colonies than stages of development and of regression. The yield was higher by selecting the CL's periphery, which might be due to capillary sprouts originating from the former thecal cell layer. For type 5, the CL of development is recommended for the selection of a high number of colonies.

5.2.1 Effects of IFN-γ on CK$^+$ Cells from the CL Compared to CK$^-$ Cells and to Surface Epithelial Cells

Interferon-γ belongs to the most important endogenous regulators of immunoresponses, among them the big impact on DC function (Billiau and Matthys 2009; Rojas and Krishnan 2010). Interferon-γ-stimulated DCs heavily upregulate MHC II for antigen presentation to T cells. Interferon-γ seems to affect the differentiation and maturation of DCs, and IFN-γ is responsible for the DC-dependent production of IL-12. Senescence and cell death is not an IFN-γ-dependent effect on DCs, yet reported for treated endothelial cells (Friesel et al. 1987; Stolpen et al. 1986). For this reason, our own findings on IFN-γ-treated CK$^+$ cells in comparison with the other phenotypes become important.

In flow cytometric analysis, IFN-γ (200 U/0.5 ml for 3 days) was a more potent inducer of MHC II antigens in confluent cultures of types 2 and 5 than in types 1, 3 and 4 (Spanel-Borowski and Bein 1993). The IFN-γ-induced de novo expression of MHC II products in type 5 strikingly, which was about 80-fold versus untreated cells (Fig. 5.13; Table 5.1). Cells of types 1 and 2 additionally demonstrated a high unspecific binding of immunoglobulins indicating a substantial number of Fc-binding sites. The growth rate of the five phenotypes was then studied starting at a low cell density with pulse treatment of 200 or 1,000 U/0.5 ml IFN-γ for 3 days. Cell number was determined with an electronic particle counter on days 4, 7, 10 and 13. In the absence of IFN-γ, the growth rate was high for the types 3 and 4, moderate for type 1, and low for types 2 and 5 (Fig. 5.14). The IFN-γ treatment exerted anti-proliferative effects in types 1–4, and had the tendency to support cell growth of type 5. It is obvious that not type 1 but type 5 displays an IFN-γ-related signature with strong MHC II peptide upregulation and proliferative activity, which is in line with the IFN-γ-related DC response (Banchereau and Steinman 1998; Billiau and Matthys 2009). In contrast, many Fc-binding sites characterize type 1 cells, which is typical for DCs.

Treating confluent monolayers with 200 U/0.5 ml IFN-γ for 3 days, the cellular response was cell-type specific (Fig. 5.15). Cell cohesion persisted in types 1 and 2 accompanied with flattening of the monolayer, whereas the integrity of the types 3

Fig. 5.9 (continued) abundant intermediate filaments (*lower right corner*) contains granules of low and high electron density. High-dense granules could be lysosomes, because a high-dense granule is fusing with a low-dense granule (*arrow* in **e**). (**f**) A classical desmosome with an intermediate dense line connects two cells. Adapted from Spanel-Borowski (1991), Spanel-Borowski and van der Bosch (1990) and Fenyves et al. (1994)

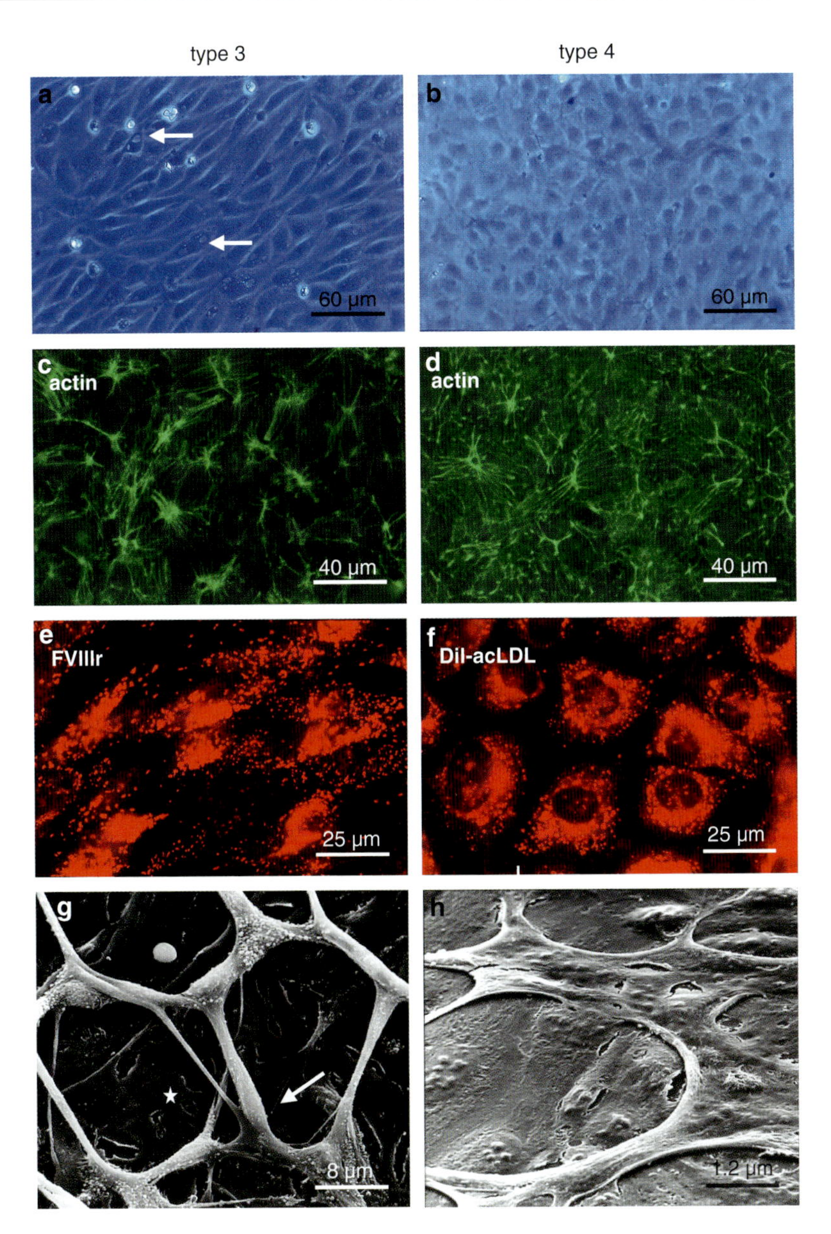

Fig. 5.10 Endothelial cell criteria in cell types 3 (*left*) and type 4 (*right*) cultures from the bovine CL under the phase contrast microscope (**a** and **b**), by staining with phalloidin-FITC after 20-min fixation with 2% paraformaldehyde containing 0.2% Triton-X for permeabilization (**b** and **c**), by immunofluorescence localization after fixation with absolute ice-cold ethanol for 5 min (**e**), by uptake of Dil-acLDL (10 μg/ml in serum-containing medium at 37°C for 4 h in **f**), and for the study of ultrastructure after fixation with 2% glutaraldehyde for 1 h followed by post-fixation with 2% osmium tetroxide (**g** and **h**) (**a** and **b**) The type-3

and 4 monolayer was loosening and senescent cells with large-sized vacuoles appeared (Fenyves et al. 1993). Similar reports on antiproliferative IFN-γ-related effects have been published for endothelial cells (Friesel et al. 1987; Stolpen et al. 1986). The treated type 5 monolayer changed into a multilayer because of cell proliferation. Intercellular junctions conspicuously developed in type 1 cells under IFN-γ treatment (Fenyves et al. 1993; Ricken et al. 1996). The contact reinforcement transferred to augmented expression of tight junction proteins such as ZO-1 and occludin. Tight junction strands were increased in freeze fracture replica (Fig. 5.16a, b, g, h), and gap junctions decreased. The paracellular space was completely blocked impeding the passage of horseradish peroxidase (HRP; Ricken et al. 1996). Vinculin, plakogobin, desmoplakin, desmoglein and E-cadherin for zonulae adherentes and desmosomes were also upregulated in the IFN-γ-treated type 1 cultures (Fig. 5.16c–f). Noteworthy was that cell type 5 showed the increased expression of N-cadherin as a linear and broken structure in immunofluorescence staining, which might be the site of filipodia (Fig. 5.15i, j; compare with filipodia in Fig. 5.11; Fenyves et al. 1993). Taken together, the presence of tight junctions in type 1 cells alludes to the tight junction protein expression in DCs to generate micro-compartments in the gut epithelium (Rescigno et al. 2001). The reinforcement of tight junctions, desmosomes and adherens junctions is little known as an IFN-γ-dependent effect. The increased N-cadherin expression in cytokine-treated type 5 cells probably exerts another function in considering the IFN-γ-dependent cell proliferation (Fig. 5.14).

The steroidogenic CK$^+$ cells appear to have originated from the sex cord cells (Fig. 4.1), and the microvascular CK$^+$ cells from the sex cord region in the early fetal period (see Sect. 5.2.2). It is thus conceivable that similarities exist between CK$^+$ cells and the cells from the surface epithelium, which represent the progeny of sex cord cells. Cells were obtained by gently scraping the surface of bovine ovaries with a cotton-wool swab (Auersperg et al. 2001). Cells were released by medium flush and cultured like cells isolated from the bovine CL. Confluent cultures looked like type 1 cells and expressed a network of CK filaments

Fig. 5.10 (continued) monolayer contains spindle-shaped endothelial cells with single intracellular vacuoles (*arrow* in **a**), whereas type 4 cells are round cells and without vacuoles (**b**); the monolayer of flat cobble-stone structure looks “blurred” because of fibrin filaments on the apical side (not shown). (**c** and **d**) The actin filaments form prominent star-like nodal points in type 3, and a regular network with tiny nodal points in type 4. (**e** and **f**) Types 3 and 4 display granules with FVIIIr antigen and lysosomes with Dil-acLDL. (**g** and **h**) Tubules/pseudotubules are depicted by the scanning electron microscope cells after dehydration of the specimens with the critical-point drying method and coating with palladium. The 3-dimensional tubular network of type 3 cultures lacks the borders of the composing cells, yet nuclei can be localized (*arrow* in **g**). The surface of the monolayer is smooth (*asterisk* in **g**). A pseudotubular network occurs in type 4 cultures because of the 2-dimensional aspect (**h**). Adapted from Spanel-Borowski (1991), Spanel-Borowski and van der Bosch (1990) and Fenyves et al. (1994)

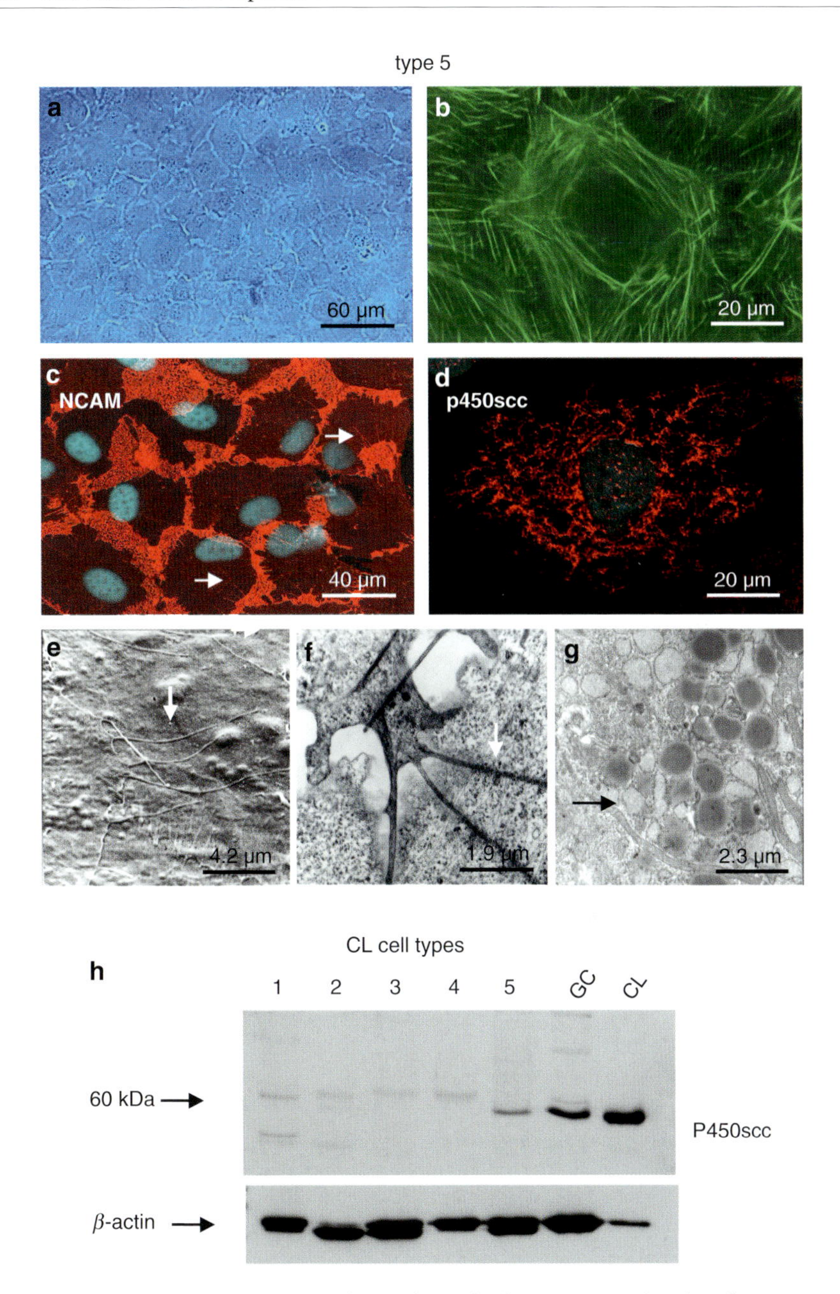

Fig. 5.11 Steroidogenic cell type 5 cultures from the bovine CL under the phase contrast microscope (**a**) for staining with phalloidin-FITC after 20-min fixation with 2% paraformaldehyde containing 0.2% Triton-X for permeabilization (**b**), for immunofluorescence localization after fixation with absolute ice-cold ethanol for 5 min (**c** and **d**), for the study of ultrastructure after fixation with 2% glutaraldehyde for 1 h followed by post-fixation with

(Fig. 5.17a, b). The 3-day treatment with 200 U/ml IFN-γ increased E-cadherin-positive adherens junctions, which were regularly distributed at the lateral cell side and at cell corners (Fig. 5.17c, d). Completely closing tight junctions developed, because neglectable HRP had passed the intercellular space of treated cultures (Fig. 5.17e). In comparison, untreated cultures allowed enzyme passage between 1 and 3 h of exposure. It is amazing that the CK$^+$ cells isolated from the bovine CL maintain the features of their far-distant parent tissue (Fenyves et al. 1993; Auersperg et al. 2001). Both have apical microvilli and a single cilium (Fig. 5.7b, c, d; Nicosia and Johnson 1984). CK$^+$ cells form tight junctions, simple desmosomes, and cadherins (Figs. 5.6h, 5.7a and 5.17c, d), as does the surface epithelium. Both cell types express vimentin and CK. A changing phenotype, being either epithelioid or fibroblast-like, is characteristic for cells of the surface epithelium. The fibroblast-like CK$^+$ cells are demonstrated in regressing antral follicles (Fig. 4.2f, g) and appear at the time of follicle transformation into a CL (Fig. 4.3a). The fibroblast-like CK$^+$ type is obtained from follicle aspirates (Ben-Ze'ev and Amsterdam 1989; Löffler et al. 2000) and from cumulus oophorus complexes (Fig. 5.5; Serke et al. 2010). The fibroblast-like type could result from an epithelio-mesenchymal conversion, which is induced in the surface epithelium at the site of follicle rupture by epidermal growth factor, TGF-β, ascorbate and collagen fragments (Auersperg et al. 2001).

5.2.2 Reflections on Quality of the CK$^+$ Type 1 and Similarity with the Type 5

At present, it is not possible to correctly pursue the fate of CK$^+$ cells from the fetal to the adult ovary as well as from the preovulatory follicle into the CL. As derived from the major population of preovulatory CK$^+$ granulosa cells, most of them seemed to transform into the frequently occurring steroidogenic CK$^+$ cells in the early CL (Fig. 4.3). The large ones contained neurophysin, whereas the small-sized CK$^+$ cells were without the neuropeptide (Table 4.1; Ricken et al. 1996). A minor population of CK$^+$ cells was integrated into microvessels of the CL parenchyma

Fig. 5.11 (continued) 2% osmium tetroxide (**e–g**), and for the western blot analysis of cytochrome P450scc (**h**). (**a**) The flat monolayer seems to represent contact-inhibited growth. (**b**) Actin filaments are arranged as a prominent circular band. (**c**) The cell margin strongly expresses NCAM-140. (**d**) The P450scc-positive network corresponds to the steroidogenic machinery in the smooth endoplasmic reticulum. (**e** and **f**) The scanning and transmission microscope reveals extremely long filipodia (*arrow*). (**g**) The cytoplasm depicts lipid droplets and a dilated smooth endoplasmic reticulum (*arrow*). (**h**) The immunoband for P450scc is positive for type 5 cells, not for types 1–4 in the representative western blot. Positive controls are lysates of granulosa cells (GC) and of the bovine CL. Adapted from Spanel-Borowski and van der Bosch (1990), Spanel-Borowski (1991), Mayerhofer et al. (1992) and Fenyves et al. (1993)

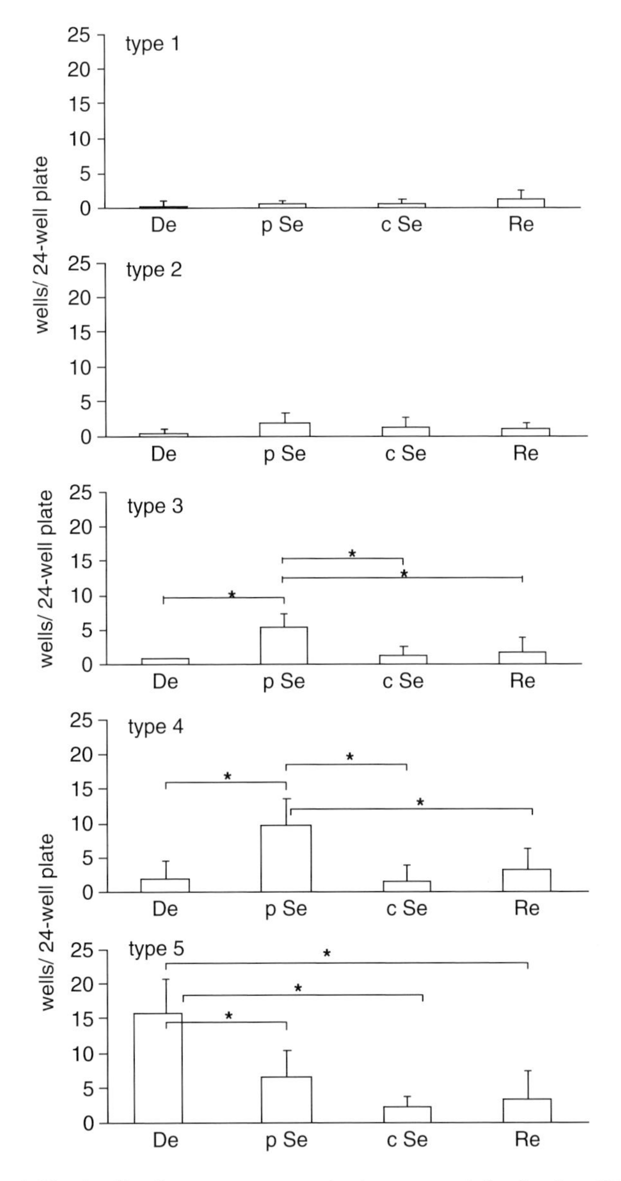

Fig. 5.12 The yield of cell cultures types 1–5 is documented for bovine CL at stages of development, secretion, and regression (*De, Se, Re*) by taking into consideration the peripheral and central region in the CL of secretion (*pSe, cSe*). The isolation and cultivation of the five phenotypes is described in Table 2.1. Positive wells from a 24-well plate are those which show prominent colonies that allow colony transfer (here given as *well/24-well plate*). The colony yield remains constant for types 1 and 2 throughout the estrous cycle. The output for types 3 and 4 is significantly higher for the peripheral than the central portion of the CL at the stage of secretion. Type 5 recovery is the highest at the stage of development, $*p < 0.05$

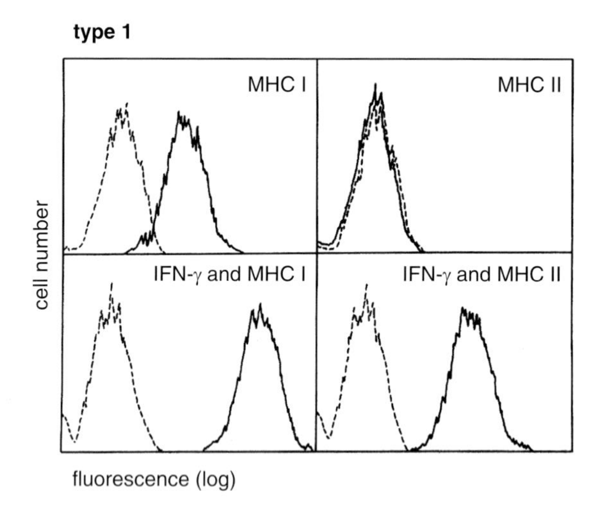

Fig. 5.13 The MHC I and II antigen expression is shown as a graph for cell type 1 after a 3-day exposure with IFN-γ (mean fluorescence intensity for types 1–5 in Table 5.1). The 1×10^6 FITC-labeled cells were analyzed per untreated or treated sample with Coulter Epics Profile II System. Each histogram is related to 5,000 cells (*Y*-axis). Unspecific staining as *broken line* and fluorescence intensity as fluorescence (*X*-axis). Adapted from Spanel-Borowski and Bein (1993)

Table 5.1 Mean fluorescence intensity[a] on five phenotypes from bovine CL

phenotypes/treatment	1	2	3	4	5
	Unspecific binding				
None	8[b]	6[b]	2	6	3
	3	2	0	4	1
IFN-γ	18	23	2	7	3
	5	6	0	5	0
	MHC I				
None	35[b]	47	119	114[b]	84[b]
	5	13	19	16	26
IFN-γ	457	345	153	283	584
	84	52	10	42	96
	MHC II				
None	6[b]	6[b]	11[b]	7[b]	3[b]
	2	1	4	3	1
IFN-γ	48	163	50	50	233
	7	27	14	18	66

Adapted from Spanel-Borowski and Bein (1993)

[a]The geometric mean peak fluorescence intensity of MHC I and II antigen expression was studied with fluorescence flow cytometry. The intensity was calculated by integrating the area under the histograms for type 1 (Fig. 5.13). The area corresponding to unspecific binding was subtracted. Data are means ± SE of 5–8 independent assays run for each cell type by using a different cell line.

[b]$p < 0.02–0.05$ untreated versus IFN-γ

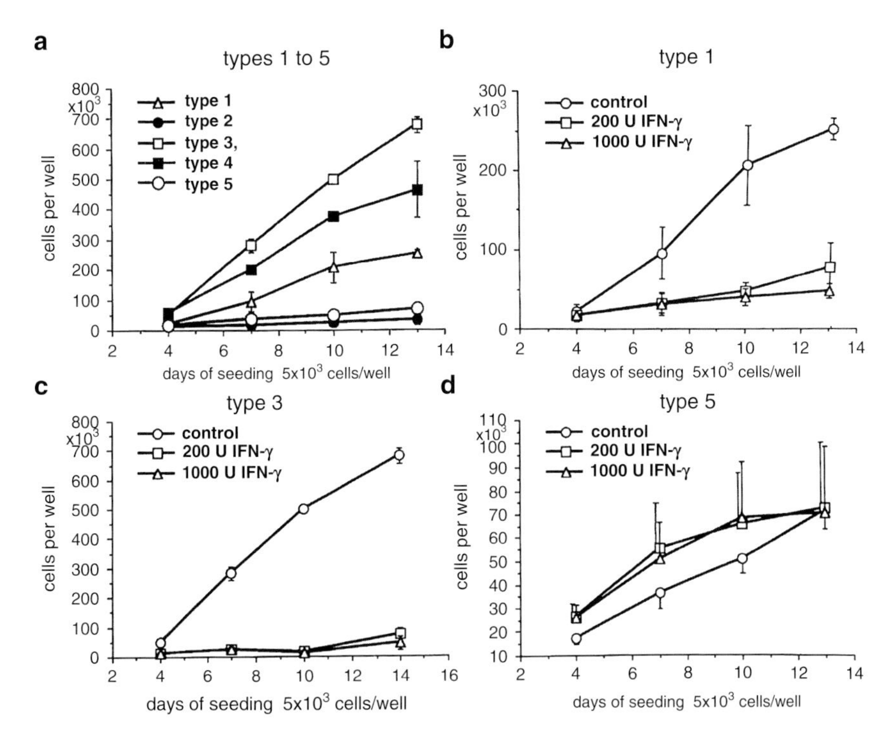

Fig. 5.14 Different growth response curves of untreated (types 1–5, **a**) and IFN-γ-treated cell types 1, 3 and 5 (**b–d**). On day 0, 5×10^3 cells were seeded per well of replicate 24-well culture plates. After buffer wash the next day, cultures were untreated or treated with 200–1,000 IU of recombinant bovine IFN-γ/0.5 ml culture medium on day 1 for 3 days. On days 4, 7, 10 and 13, cells were enzymatically dislodged and counted with an electronic particle counter. Absolute numbers were calculated for each cell type as well as the mean ± SD from two independent experiments. The basic growth rate is high for types 3 and 4, moderate for type 1, and low for types 2 and 5. Interferon-γ clearly acts anti-proliferatively in type 1 and 3 cultures, whereas cell growth is not inhibited in type 5 cells. Adapted from Fenyves et al. (1994)

and septum (Fig. 4.3f–h) and thus represents a microvascular CK$^+$ cell. The microvascular CK$^+$ cell type in the intact CL could correlate with the CK$^+$ type 1 (and also type 2) derived from the CL (Table 5.2). Type 1 showed contact-inhibited growth, FVIIIr antigen expression, and uptake of Dil-acLDL similar to endothelial cells (Fig. 5.6a, d, e). A difference related to the perinuclear diffuse FVIIIr antigen response in type 1 cells, whereas endothelial types 3 and 4 displayed the characteristic granules (Figs. 5.6d and 5.10e). The steroidogenic enzyme P450scc was missing in type 1 (Fig. 5.11h). It proliferated in culture (Fig. 5.14), which would not be expected for a highly differentiated steroidogenic cell. Another reason to judge CK$^+$ type 1 cells as microvascular cells was their angiogenic potential as revealed by semiquantitative and competitive RT-PCR analysis (Tscheudschilsuren et al. 2002).

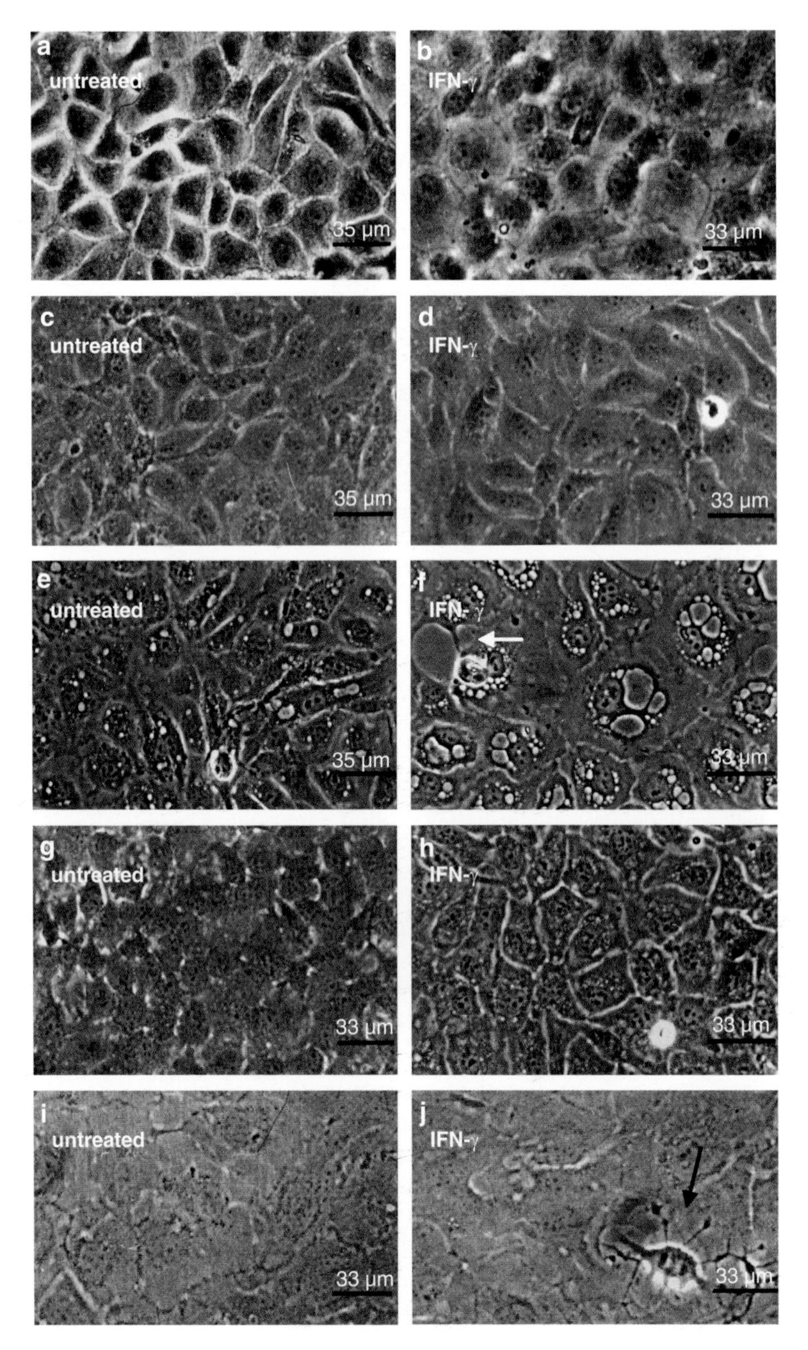

Fig. 5.15 Different effects of IFN-γ on confluent cultures of types 1–5 derived from the bovine CL. Cultures were untreated (*left*) or treated once with 200 IU of recombinant

The CK^+ cells had mRNA for VEGF-2 receptor (fms-tyrosine kinase-1 receptor), and for the Tie-2 receptor binding angiopoietin 1 and 2. No mRNA for VEGF-1 receptor (kinase-insert domain-containing receptor) or for angiopoietin 1 and 2 were seen under hypoxic conditions. The qualitative and quantitative validation by flow cytometric analysis disclosed further endothelial cell criteria for the microvascular CK^+ type 1 (Lehmann et al. 2000). The possible interaction between endothelial cells and the extracellular matrix is given, because molecules such as CD29, CD31, CD49a and CD62P were expressed although at lower levels than in types 3 and 4. The expression difference in these proteins was maintained under a 12-h treatment with 0.5 and 5 ng/ml TNF-α. In contrast to the positive IFN-γ effect, the MHC II peptides remained low in resting CK^+ cells, and in TNF-α-stimulated cells. Most importantly, the yield of type 1 and 2 colonies established from CL remained alike throughout the estrous cycle, which was in contrast to the types 3–5 (Fig. 5.12). It signifies that a minority of the microvascular CK^+ cell need no gonadotrophins for survival. Indeed, cell type 1 lacked the receptor mRNA for FSH and LH, whereas prostaglandin F_{2a} and E_2 (PGF_{2a} and PGE_2) receptor mRNA were seen (Aust et al. 1999).

The fact is that CK^+ endothelial cells are present in quite different organs without knowing their function and origin (Jahn et al. 1987; Mineau-Hanschke et al. 1993; Patton et al. 1990). Our view is that the microvascular CK^+ cells have originated from the aorta-gonado-mesonephric region far back in fetal life (Dieterlen-Lièvre et al. 2006; Pouget et al. 2006). As is shown by the genetic tracing strategy for VE-cadherin expression in fetal mice, somites release angioblasts with endothelial and hematopoietic quality (Zovein et al. 2008). The angioblasts are located by VE-cadherin positivity first in the aorta, from where they migrate to the liver and later to the bone marrow. Single VE-cadherin-positive cells populate microvessels of different adult organs, also of the uterus. They might also be displayed in the CL and revealed as CK^+ cells with a close look. The so far overlooked aspect is that the angioblasts in the vicinity of the primitive aorta and the mesonephric duct reside in the intermediate mesoderm of the splanchnopleura (Dieterlen-Lièvre et al. 2006), and are limited by the thickened coelomic epithelium of the later genital ridge. Thus, the intimate anatomical relationship allows a very early temporary contact between the angioblasts and the forming primitive gonad. The possibility is given that the adult aorta also hosts an

Fig. 5.15 (continued) bovine IFN-γ/0.5 ml (per well of a 24-well plate) for 3 days (*right*) and studied under the phase contrast microscope. (**a** and **b**) The treated type 1 monolayer flattens and loses the epithelioid cobble-stone appearance in comparison with the untreated culture. (**c** and **d**) The IFN-γ-related response of the type 2 culture is similar to type 1 changes. (**e** and **f**) The treated type 3 monolayer consists of enlarged, senescent cells with striking vacuoles. The intercellular space widens (*arrow*). (**g** and **h**) The IFN-γ-treated type 4 monolayer shows polygonal and flattened cells. (**i** and **j**) A cell in IFN-γ-treated type 5 monolayer is dividing (*arrow*). Adapted from Fenyves et al (1993)

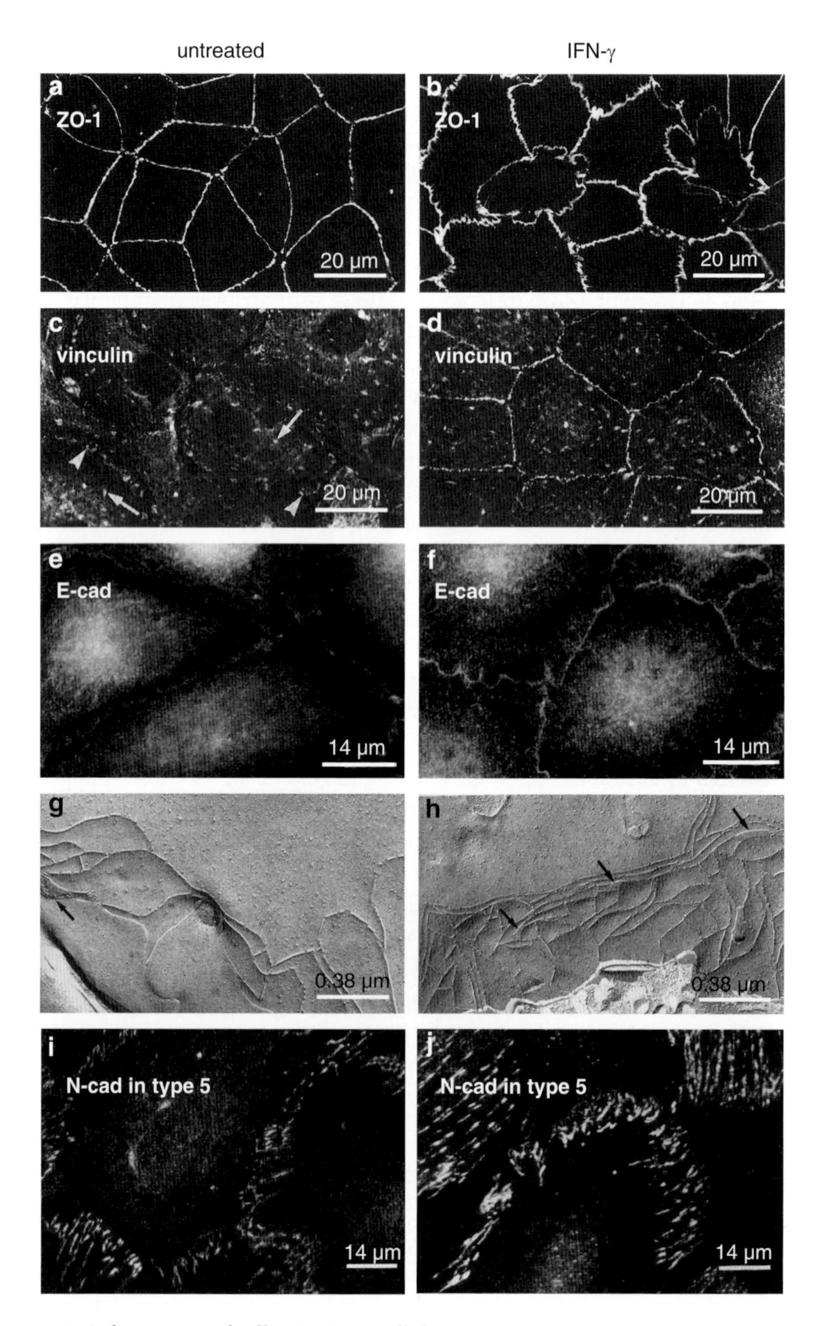

Fig. 5.16 Reinforcement of adhesive intercellular contacts in IFN-γ-treated type 1 monolayer (*right*) in comparison with untreated cultures (*left*) using immunofluorescence localization after fixation with ice-cold absolute ethanol for 5 min (**a–f**) and the freeze fracture technique for cells fixed with 2.5% glutaraldehyde (**g** and **h**). Type 5 cell is also

angioblast population with CK expression, because a vascular CK^+ cell type was isolated from the bovine aorta (Spanel-Borowski et al. 1994b), and cells with a single cilium were casually found in the intact bovine aorta (unpublished observation with the scanning electron microscope). In considering the single cilium as the cell's antenna (Davis et al. 2006; Singla and Reiter 2006), the vision is favored that the cilium of the microvascular type 1 cell (Fig. 5.7b–d) senses danger signals such as oxidative stress and alarmins. Additional support comes from novel ciliary protein function, which is being implicated in Wnt-signaling, and, in turn, Wnt-signaling could fine-tune the TLR pathway (Gordon et al. 2005; Blumenthal et al. 2006). Taken together, the microvascular CK^+ cells might reside as immune guardian in a lumen's wall, there being the antenna of the vascular system. The microvascular CK^+ cells are suggested to have a far distant progeny from the aorta-gonado-mesonephric region. They are separate from KIT-positive endothelial precursor cells, which could be harvested from bovine follicles and likely from the bone marrow (Merkwitz et al. 2010).

Why 26 of 45 bovine CL depicted CK^+ cells, and 7 CL from pregnant cows lacked any CK^+ cells remains unclear (Ricken et al. 1995). A loss of CK gene expression might be the cause as is described for transformed cells due to the rapidly degraded CK18 protein in the absence of the complex partner CK8 (Knapp and Franke 1989). Such an event could also have occurred during transition from the primordial follicle into the preantral follicle. The appearance of CK^+ granulosa cells in preovulatory follicles and in regressing follicles (Fig. 4.3) indicates that not LH, but specific intra-ovarian factors might have reactivated the CK 8 gene. Provided it is switched off after ovulation, the steroidogenic CK^+ cell gradually becomes a CK^- granulosa-like cell subsequently characterized as type 5 (Spanel-Borowski et al. 1994a). The postulated conversion into CK^- granulosa-like cell is mainly attributed to the small type (20 μm in diameter), which dominated the morphological scenery in the developing CL (87% of small cells vs. 13% of large cells above 40 μm in diameter; Fig. 4.3c, d; Ricken et al. 1995). A transition of small cells into large CK^+ cells is largely excluded, because the number of CK^+ cells per field decreased when comparing stages of development and secretion (74 ± 6 and 25 ± 6; Ricken and Spanel-Borowski 1996). A delayed CK switch-off could

Fig. 5.16 (continued) studied (untreated and treated in **i** and **j**). (**a** and **b**) The ZO-1 antigen, which occurs at the cytoplasmic side of tight junctions, is intensified in expression in treated cells. (**c** and **d**) The vinculin molecule binds actin filaments of adherens junctions, and it is upregulated in treated cultures (*arrowheads* indicate the lateral cell border and *arrows* the cell surface) (**e** and **f**) The E-cadherin (E-cad) antigen, a Ca^{2+}-dependent transmembraneous glycoprotein, contributes to adherens junctions and desmosomes; its expression is increased in the IFN-γ-treated monolayer. (**g** and **h**) The freeze fracture replica reveals a complex tight junction with 7 strands in the treated culture compared to 3 strands in the control. (**i** and **j**) N-cadherin (*N-cad*) is increased at the filipodia of treated type 5 cells; see filipodia in Fig. 5.11e, f. Adapted from Fenyves et al. (1993) and Ricken et al. (1996)

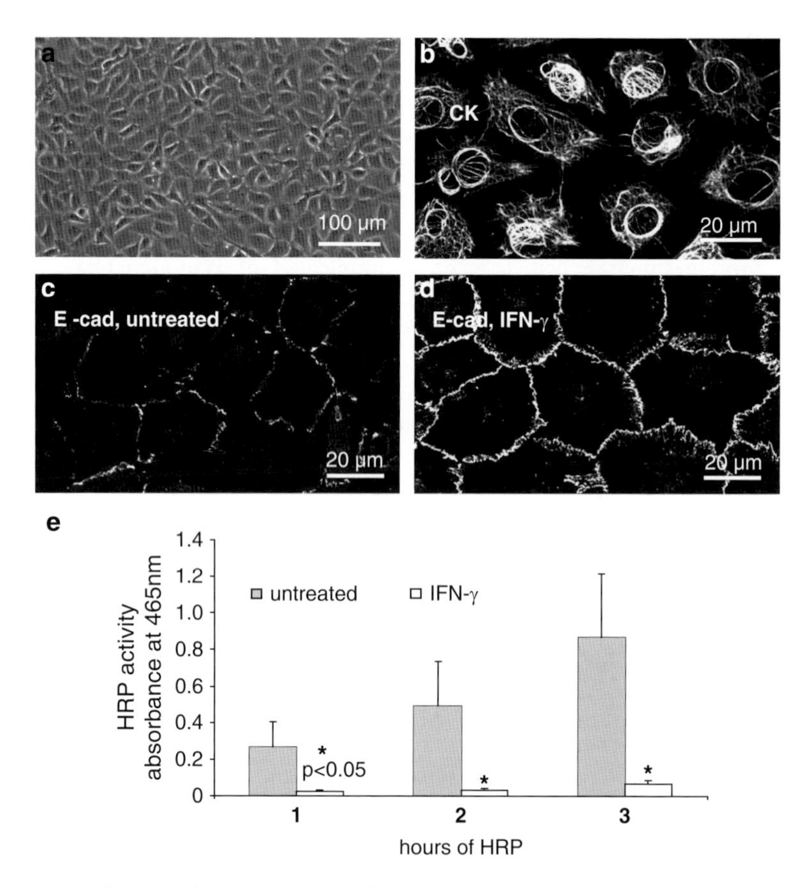

Fig. 5.17 Similarities of the surface epithelium cell culture compared with type 1. The surface epithelium of the bovine ovary was gently scraped with a cotton swab, and the swab twisted in serum-containing culture medium to release the cells; confluent cultures were studied with the phase contrast microscope (**a**) and with immunofluorescence localization after fixation with ice-cold absolute ethanol for 5 min (**b**–**d**). Cultures were untreated and treated with 200 IU of recombinant bovine IFN-γ/0.5 ml culture medium for 3 days (**c** and **d**). For the permeability study, confluent cultures were raised on the membrane of 0.45 μm^2 pore size in the upper compartment of a 24-well cluster plate, and 0.2 mg/ml HRP added after IFN-γ exposure. (**h**) Samples were taken from the lower compartment at hours 1, 2 and 3 to determine HRP activity with the 3,3′-diaminobenzidine reaction in a spectrophotometer. Data are mean ± SD from 3 independent experiments. (**a**) The epithelioid monolayer of cobble-stone pattern. (**b**) All cells produce a dense network of CK. (**c** and **d**) E-cadherin (*E-cad*) is upregulated at the cell border in the treated culture. (**h**) The statistically significant increase in HRP activity in the lower compartment gives clear evidence for impermeable intercellular contacts. Adapted from Spanel-Borowski et al. (1994) and Ricken et al. (1996)

explain the varying intensities of CK expression in steroidogenic CK$^+$ cells as demonstrated by immunostaining in the CL of secretory stage and of regression (Fig. 4.3d, e). The switch-off seems to be completed in CL of pregnancy that misses

Table 5.2 Findings for CK^+ type 1 and for type 5 isolated from the bovine corpus luteum

Findings	Type 1	Type 5	References
Ultrastructure	Long single cilium, microvilli, granules	Ciliary stub, long filipodia, lipid droplets	Spanel-Borowski and van der Bosch (1990), Spanel-Borowski (1991) and Wolf and Spanel-Borowski (1992)
Intercellular contact molecules	NCAM at adhesion plates, Fig. 5.7a	NCAM, lateral cell side, Fig. 5.11c	Mayerhofer et al. (1992)
Cytoskeleton	Peripheral band of actin, network of cytokeratin	Circular band of actin	Fenyves et al. (1993)
Endothelial cell criteria	Diffuse FVIIIr response, uptake of acLDL α2-macroglobulin	Diffuse FVIIIr response uptake of Dil-acLDL	Spanel-Borowski and van der Bosch (1990)
Angiogenic mRNA	VEGF, Flt-1 TIE-2	Not done	Tscheudschilsuren et al. (2002)
Steroidogenic enzymes	Absent	3β-HSD P450scc	Spanel-Borowski et al. (1994) Fig. 5.11h
mRNA receptors	Absent for FSH and LH Present for PGE_2 and PGF_{2a}	Not done	Aust et al. (1999)
Phorbol myristate acetate stimulation	Leucocyte adhesion up by β2 integrin (CD18)	Leucocyte adhesion up by CD18/β2 integrin	Ley et al. (1992)
TNF-α stimulation 0.5–5 ng/ml for 12 h	mRNA:M-CSF up, GM-CSF absent, proteins up: CD29, CD31, CD49a, CD62P, CD49b absent	Not done	Lehmann et al. (2000)
KIT	Not done	48-kDa fragment	Koch et al. (2009)
Interferon-γ (200 U/ml for 3 days)			
Cell growth	Anti-proliferative, reinforced monolayer	Proliferative multilayer with loss of contact inhibition	Fenyves et al. (1994) and Ricken et al. (1996)
Zonula adherens, desmosome	E-cadherin up, Vinculin up Plakoglobin up Desmoglein up Connexion 43 down	N-cadherin up	Ricken et al. (1996) Fenyves et al. (1993)
Tight junction	ZO-1 up	Absent	Ricken et al. (1996)
MHC II versus untreated cells	Sevenfold	80-fold	Spanel-Borowski and Bein (1993)
Unspecific IgG binding	Threefold	None	Spanel-Borowski and Bein (1993)

CK-positive cells (Ricken and Spanel-Borowski 1996). When our thoughts prove to be correct, the CK^{-} granulosa-like cells obtain another outfit to modify the potential immune function of steroidogenic CK^{+} cells. The conspicuous filipodia of type 5 (Fig. 5.11e, f) would well fit dendrites of mature DCs (Banchereau and Steinman 1998). The N-cadherin-positive filipodia (Fig. 5.16i, j) could be the site of MHC II antigen upregulation under IFN-γ stimulation and thus of antigen presentation to interact with lymphocytes. The function of type 5 cells might be similar to mature DCs, which recruit and activate T cells (Banchereau and Steinman 1998). It is presently a big challenge for immunologists to prove that the organ itself is in control of danger and the restoration of tissue integrity (Matzinger 2007). The author provides exciting reflections that the ultimate power of immunosurveillance lies within the organ itself and assumes that organ-specific demands have tailored unique effector cells. It is being learnt that professional DCs derived from the bone marrow and from blood monocytes are not the whole story and that an organ cares for its own protection device. In the epidermis, Langerhans cells develop from an embryonic precursor cell before birth, and establish a network during the first postnatal weeks (Chorro et al. 2009). In the ovary, the steroidogenic CK^{+} cells, which have originated from the surface epithelium in the early fetal life, come into focus as organ-tailored DCs. The surface epithelium and the sex cords as offsprings have lived together with germ cells (Fig. 4.1), which might have imprinted genes of the sex cord cells for an immunological message in the adult period. Not all granulosa cells have their progeny from the sex cord cells. A subset might come through fusion with the CK^{+} cells from the mesonephric tubules (Fig. 4.1b) and deliver the working class of steroidogenic cells. Of note, in the adult ovary, more than 85% of human ovarian cancers arise from the surface epithelium (Auersperg et al. 2001) or from dysplastic epithelial cells shed from the Fallopian tube and successfully implanted into the surface epithelium (Kuman and Shih 2010).

In context with the possible origin of microvascular CK^{+} cells in the vicinity of the genital ridge (see above), similarities between types 1 and 5 are of interest (Table 5.2): a diffuse perinuclear FVIIIr response, a circular arrangement of actin filaments, NCAM-140 expression at the lateral cell side, the single cilium either as long cilium in type 1 (Fig. 5.7b–e) or as ciliary stub (Wolf and Spanel-Borowski 1992), and a comparable high leucocyte adhesion mediated by the β2 (CD18) integrin after unspecific stimulation with phorbol myristate acetate (Ley et al. 1992). The percentage of adhesion significantly surmounted the leucocyte adhesion in types 2–4. Type 5 does not produce full-length KIT of 160–135 kDa, a reliable marker of precursor cells, but a small 48-kDa fragment with unknown function (Koch et al. 2009). Full-length KIT is generated by thecal cells, which become the theca-derived small luteal cells in the developing CL (Spanel-Borowski et al. 2007). The completely unclear molecular interaction between the steroidogenic CK^{+} type from the former granulosa (Fig. 4.3a, b) and the KIT-positive thecal cells adds to the complexity of the CL world.

Chapter 6
Working Hypothesis and Challenges

Follicle rupture and CL formation can be compared with the breakdown of a skyscraper and its subsequent reconstruction with new architecture. Thus, an ingenious machinery is at work to manage a controlled tissue damage and its recovery for the next ovarian cycle. The INIM arises on the backs of the endocrine system. We suggest that the breakdown process is triggered by oxidative stress from the follicles inside, thus the signaling cascade goes from the granulosa towards the theca (Fig. 6.1). The process starts with ROS as by-products of steroidogenesis (Derouet-Humbert et al. 2005; Hanukoglu 2006; Yacobi et al. 2007), which is at its maximum in preovulatory follicles under the LH pulse. ROS is released into the follicular antrum and oxidizes nLDL to oxLDL. The oxLDL-dependent activation of LOX-1 in CK^{-} granulosa cells adds more ROS by a vicious internal feedback (Dandapat et al. 2007; Mehta et al. 2006). The preovulatory follicle turns into a structure under oxidative stress, which is in line with the findings and conclusions by others (Agarwal et al. 2003, 2005). At the culminating point, sensitive granulosa cells become damaged and release alarmins (Bianchi 2007; Rock et al. 2010). In particular, the blebs of apoptotic cells are reported to be enriched with oxidized lipids (Hartvigsen et al. 2009), which are recognized by PRRs. Our working hypothesis has a solid background, because LOX-1 and TLR4, which are receptors for oxLDL, have been found in human granulosa cells and oxLDL in the follicle fluid (Bausenwein et al. 2010; Duerrschmidt et al. 2006; Serke et al. 2009). In addition, 20–50% of dead cells were counted in fresh follicle harvests (Vilser et al. 2010). The alarmins from damaged cells could then mediate the release of soluble complement factors (C1q, C3a, C5a) as rapid danger sensors and transmitters of oxidative stress: the factors interact with specific complement membrane receptors (Köhl 2006a, b). Early outside signaling (from granulosa towards theca) within 12 h preferentially leads to the recruitment of neutrophils and macrophages, and to the breakdown of the follicle wall and oocyte expulsion (Fig. 3.3; Brännström and Enskog 2002). Early signaling is associated with IL-1β, IL-6 and TNF-α release into the follicular fluid similar to the cytokine profile from spleen cells in a thermal injury mouse model (Brännström et al. 1994a; Paterson et al. 2003). Late inside signaling (from theca towards granulosa after oocyte release) within another 12 h commands tissue remodeling

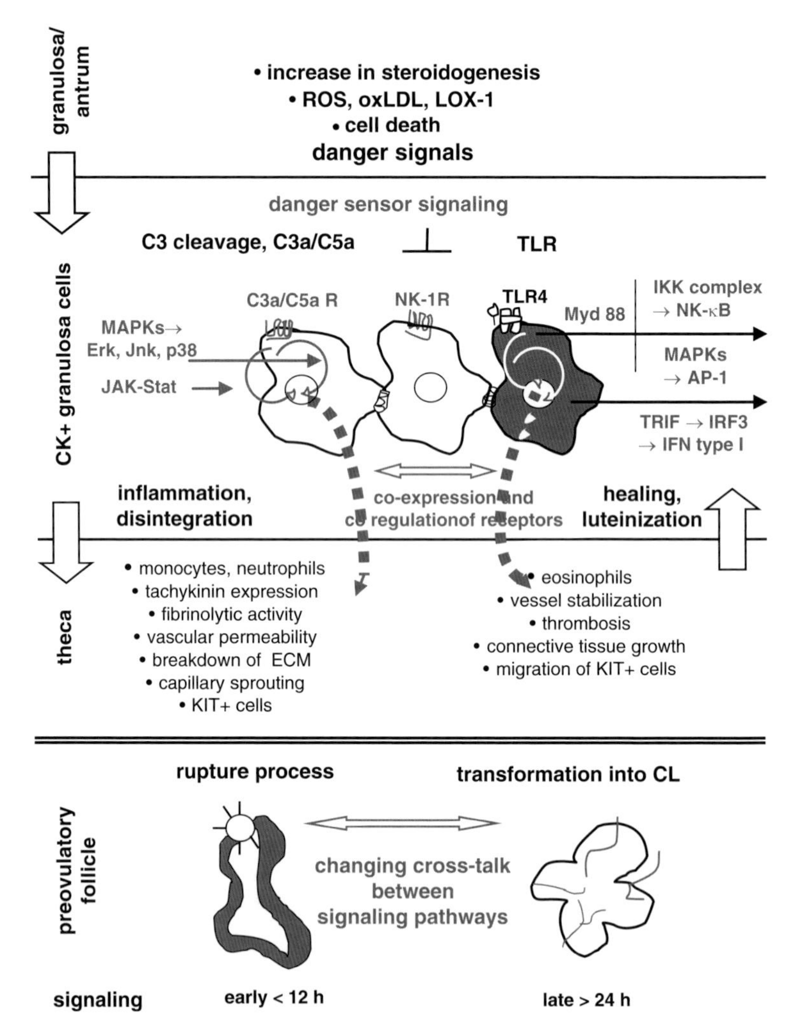

Fig. 6.1 Working hypothesis for inside-out signaling of CK$^+$ cells orchestrating follicular rupture and transformation into a CL as effector cells of INIM. The receptors for C3a and C5a as well as for tachykinin (NK-1R) could be co-expressed and co-regulated on CK$^+$ cells for the activation of the TLR4 pathway. The C3R-dependent pathways of MAPKs and of JAK-STAT lead to cell growth, differentiation cell death and pro-inflammatory events. The TLR4-related signaling activates MAPKs and IRAKs (interleukin-receptor-associated kinases) to generate immunoregulatory responses (both Myd88 dependent) through NF-κb and AP-1. The TLR4-related signaling also leads to anti-inflammatory events (TRIF-IRF3 dependent). The first sequence of interaction commands the inflammatory response with breakdown of the extracellular matrix (ECM), cell death, growth and cell differentiation under command of the AP-1 transcription factor. The second sequence governs immunoresponse by dominant NF-κB and IFR-3 transcription factors for tissue repair. Pathways for disintegration and regeneration overlap, yet the dominance shifts with time. CK$^+$ cells built a microenvironment by tight junctions. Pathways are simplified. Theoretical associations are shown by the *gray broken line*

with cell proliferation, angiogenesis, eosinophil recruitment and connective tissue repair (Figs. 3.4–3.6; DiScipio and Schraufstatter 2007). The suggested turn in functional direction could be under the influence of the KIT–KIT ligand system being responsible for cell proliferation, migration and differentiation. A dense band of KIT-positive thecal cells with full-length KIT (160–135 kDa) developed before follicle rupture. Thereafter, KIT-positive cells formed a peripheral network among the luteinizing granulosa cells (Koch et al. 2009; Spanel-Borowski et al. 2007). The reappearance of CK^+ cells in preovulatory follicles and in regressing follicles (Fig. 4.2c–g) could be due to oxidative stress under the assumption that the CK^+ cells represent danger-sensing cells. They are likely to exert this role, because the CK^+ cells regulated TLR4, CD14 and ROS under oxLDL application in culture (Figs. 5.1–5.4), which is considered as a model of oxidative stress. The complement receptor cascade might be separate from TLR4 signaling in CK^+ granulosa cells. As an alternative mechanism, the complement receptor pathway might interact with TLR4 signaling (Hajishengallis and Lambris 2010; Hawlisch and Köhl 2006). Presently, the final outcome of the molecular cross-talk between the two ancient receptor systems is beyond imagination. The TLR4 system alone confers complex signaling pathways leading to a plethora of genes in control of immunoresponses and inflammatory cytokines (Kumar et al. 2009; Takeuchi and Akira 2010). The Myd88-dependent NF-kB pathway could be responsible for the peak values of TNF-α, IL1β and of IL-6 in preovulatory follicles (Adashi 1990; Brännström et al. 1994a) and MAPKs' signaling for cell growth and differentiation during angiogenesis and luteinization. The conversation between the complement cascade and the TLR4 pathway might be relevant for outside signaling, whereas a cross-talk between the pathways of TLR and the tachykinin-regulated NK-1R (Figs. 3.4–3.6) could be responsible for inside signaling (Fig. 6.1). This suggestion is a step forward in the debate on the role of tachykinins in inflammation and ovarian functions (Debeljuk 2006; O'Connor et al. 2004). One very much wants more details about a time-dependent change in pathway conversation. The surface epithelium can be discarded as influential tissue, because regular follicle rupture occurs in epithelium-denuded primate ovaries (Wright et al. 2010) and because of our findings on IOR (Figs 3.7 and 3.8).

It is no longer under debate whether leucocytes are incidental invaders into the preovulatory follicle wall or essential effectors (Adashi 1994; Brännström and Enskog 2002; Brännström et al. 1993, 1994b). Leucocyte accumulation and the physiological wound of the follicle wall before oocyte expulsion is part of an acute and sterile inflammatory reaction, which suddenly arises under the LH surge (Espey 1994; Medzhitov 2008, 2010a). More than 30 ovulation-specific genes are involved in two crucial signaling cascades: the progesterone receptor pathway for protease production (ADAMTS-1, cathepsin L) and the EGF-family for cumulus expansion by synthesizing the hyaluron-rich matrix (Espey 2006; Richards et al. 2002; Hernandez-Gonzalez et al. 2006). That these genes are predominantly in the granulosa cell layer reflects an inside-out process, which supports our hypothesis that oxidative stress in the follicle antrum represents the primary motor of the ovulatory machinery. The inflammatory reaction of the ovulatory process seems part of an immune mechanism as first proposed (Richards et al. 2008). Comprehensive gene expression analysis of granulosa cells and cumulus cells depicted genes of the TLR family (*TLR2, 4, 8* and *9*) and the TLR adaptor molecules (*Cd14, C1q* and *Myd88*), as well as associated genes such as *CD34* and *pentraxin*. The long pentraxin3 belongs to the acute phase proteins, and acts as nodal point to stabilize the extracellular matrix of the cumulus complex (Bottazzi et al. 2006). Genes are translated to functional proteins in ovarian cells (Shimada et al. 2006, 2008). So far, these genes/proteins exclusively characterize immune cells. We here broaden the concept and consider the CK^+ granulosa cells as an ovary-specific institution to sense oxidative stress and to tailor immune responses in the preovulatory follicle and also in regressing antral follicles (Fig. 4.2e–g).

The question remains what happens with CK^+ granulosa cells after CL formation. The cells, which have orchestrated lesion and tissue repair for CL development, become unnecessary. The CL awaits another big task, which corresponds with the maintenance and then the removal of the CL. Luteolysis occurs rapidly in golden hamsters (Fig. 2.1b–e), and is delayed under the aspect of chronic inflammation in many other mammals (Fig. 2.2). The "corpus albicans", a hyalinized scar, is well known as the terminal stage of CL regression, which persists for several ovarian cycles. Our own findings and conclusions generated the concept that the steroidogenic CK^+ cells gradually switch off CK expression and become granulosa-like CK^- luteal cells, in culture termed type 5 cells (Fig. 6.2). The event might be due to changes of the endocrine micromilieu, in particular to high progesterone levels. That sex hormones govern the recruitment and function of antigen-presenting cells is shown for the female genital tract (Iijima et al. 2008). The change from steroidogenic CK^+ luteal cells to granulosa-like CK^- luteal cells remotely reminds one of immature DCs becoming mature DCs for migration to regional lymph nodes (Banchereau and Steinman 1998; Mellman and Steinman 2001). The converted cells postulated to be type 5 cells could cover different immune tasks in the CL of secretory or of regressing stages. The phenotype change provides the cellular condition to activate lymphocytes, and thus the connexion with adaptive immunity. It is remarkable that lymphocytes populate the bovine CL in increasing numbers comparing stages of development and regression (Bauer et al. 2001). When our theory holds true, type 5 represents the dock-gate to cell-mediated immune responses (Iwasaki and Medzhitov 2010).

The framework of our thoughts becomes more complex by including the presence of microvascular CK^+ cells (Figs. 4.3f–h, 5.6 and 5.7). Although they represent the minority in the intact bovine CL at the secretory stage, type 1 survives in long-term culture with maintenance of CK filaments and the formation of tight junctions and adherens junctions. As is explained under Sect. 5.2.2, type 1 could be the remote offspring of an endothelial precursor cell, which had been born in the aorta-gonado-mesonephros region (Dieterlen-Lièvre et al. 2006; Pouget et al. 2006; Zovein et al. 2008). In the embryo, it is the place where somite-derived angioblasts are generated and then migrate to various primitive

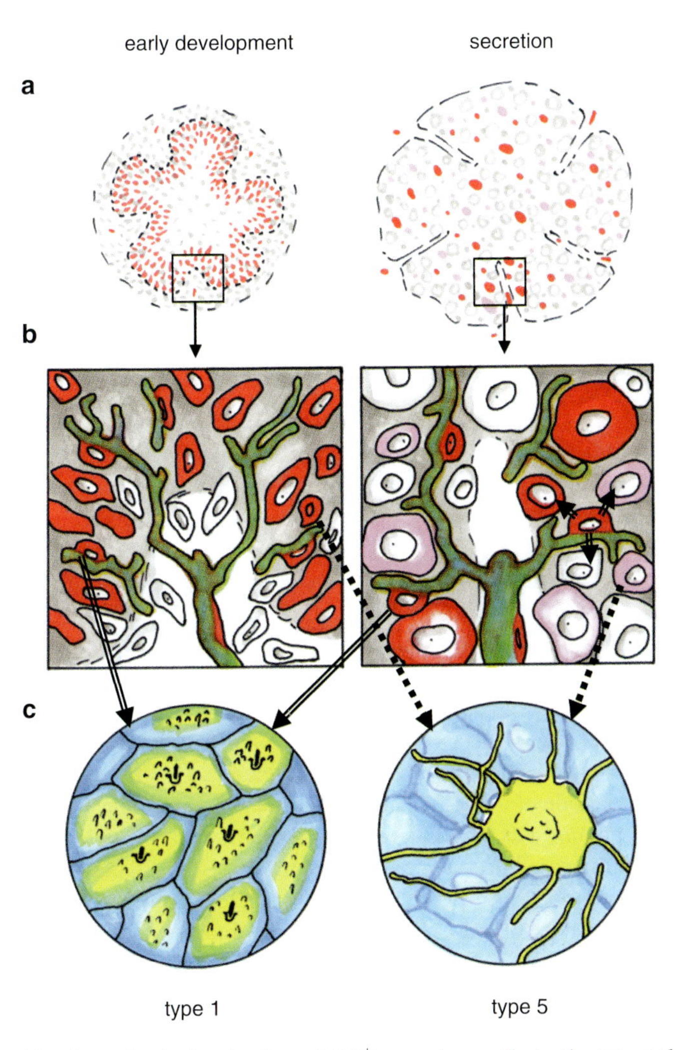

Fig. 6.2 Working hypothesis for the fate of CK$^+$ granulosa cells in the CL at the stage of development (*left*) and secretion (*right*). (a) for low magnification. At the early developmental stage, infoldings of the former granulosa cell layer generate a peripheral zone of CK$^+$ luteal cells (zonation). The former thecal layer, which forms the septum, rarely depicts a small CK$^+$ cell being an endothelial precursor cell. Zonation is absent at the stage of secretion. (b) for high magnification. At the stage of secretion, steroidogenic CK$^+$ cells of small and large size are ubiquitously distributed and demonstrate a decrease in CK intensity (red, pink and white cells) due to a switch-off in CK genes. The few CK$^+$ cells (in red), which are part of the microvessel (in green), show no difference in CK expression. (c) for cell culture. The switch-off in CK genes leads to the transition of steroidogenic CK$^+$ cells into granulosa-like cells of type 5 in culture (*broken arrows*). The microvascular CK$^+$ cells become type 1 cells in culture (*arrows*). They sense danger and signal it to steroidogenic CK$^+$ cells at the onset of luteolysis (*short open arrows*, secretory stage). Theoretical associations are *open and broken arrows in black. Drawn by R. Spanel*

organ systems and likely also to the near-by genital ridge. Such an endothelial precursor cell probably resides in the adult ovary, becomes part of the angiogenic event during CL formation (Fig. 6.2) and is seen in immunostained sections in the established microvascular bed of the CL in the secretory stage. There, the CK^+ microvascular type could sense hypoxic changes at the onset of morphological luteolysis and communicate the upcoming danger with the granulosa-like CK^- luteal cells/type 5 cells.

Although appealing because so many new aspects, we know very well that our novel concept requires a lot of work before the sovereign authority of INIM becomes generally accepted in ovarian biology. The dogma that sex hormones and intra-ovarian regulators (cytokines, chemokines and growth factors) are orchestrated by FSH and LH remains untouched. Signals of the endocrine system and of the local immune system coexist and probably interact. Our meticulous characterization of different phenotypes derived from preovulatory follicles and CL is considered as an essential basis for future experiments. The advantage is that the characterized cells are obtained from bovine and human ovaries. Cow ovaries have comparable ovarian cycles and are thus ideal models for human ovaries. Experiments should address the question whether CK^+ cells are involved in the generation and inhibition of the complement cascade in the preovulatory follicle. Details on the TLR signaling cascade and the modulation by co-regulating pathways are needed. Immunologic profiles of cytokine/chemokine secretion of surface molecule expression (MHC I and II, TLR members, and lipoprotein receptors such as CD14, CD36, and LOX-1 under TNF-α and IFN-γ treatment) might establish type 1 and 5 as immunocompetent cells. It is necessary to learn which form of cell death, whether autophagy, apoptosis or necrosis, is preferred by individual phenotypes when kept under oxidative stress such as oxLDL treatment. Such findings will broaden our understanding of follicular atresia with and without apoptotic bodies in preantral/antral follicles (Fig. 3.2; Van Wenzel et al. 1999). That survival autophagy plays a role in the maintenance of the CL has been published (Del Canto et al. 2007; Gaytn et al. 2008), yet the molecular network remains unclear. Most importantly, CK^+ granulosa cells should be experimentally converted into type 5 cells and the influence of KIT-positive thecal cells studied in culture. The in vitro findings then have to be validated with animal models, for example, by targeting the adaptor molecule Myd88. The inhibition of Myd88 in addition to the TIR domain more efficiently blocks the damage-induced acute inflammation in mice than inhibition of TLR signaling (Chen et al. 2007; Rock et al. 2010). This could be one reason why TLR4-gene deficient mice are fertile (Richards et al. 2008).

Chapter 7
Clinical Perspectives

The concept of INIM force in the biology of the ovary will generate novel strategies in the treatment of ovarian disorders. As long as ovarian INIM is balanced, it is beneficial for a controlled inflammation and a coordinated remodeling of the periovulatory events. Disorders can be understood as gain or loss of function by either overactivation or inhibition of the TLR signaling cascade. The multifold inflammatory and anti-inflammatory profiles depend on the co-regulation by other receptors and pathways (complement, tachykinin, Wnt). When the Myd88-dependent inflammatory pathway is overactivated in such a way that the high production of VEGF occurs, the hyperstimulation syndrome is generated with a life-threatening general oedema. In this context, the threat of a hyperstimulation syndrome increases with a high number of large-sized follicles at the time of oocyte aspiration (Kahnberg et al. 2009) and likely parallels a high VEGF production. Overactivation could cause the atypical follicle rupture with IOR (Figs. 3.7 and 3.8), which might be clinically hidden behind the unruptured luteinized follicle (LUF) syndrome (Qublan et al. 2006). The true LUF syndrome with high progesterone levels and the failure to become pregnant belong to anovulation disorders such as the PCOS with women suffering from androgen excess (Wild et al. 2010). A majority of women with PCOS are obese and show not only increased levels of oxLDL in the follicular fluid compared to the normal-weight counterparts but also increased levels of the oxLDL-dependent LOX-1 and up to 50% of dead granulosa cells in fresh follicle harvests (Bausenwein et al. 2010; Vilser et al. 2010). An increase in follicle danger signals could be the case in overweight women and explain why lifestyle modification with regular exercise and food restriction can restore ovulations (Rachon and Teede 2010; Thomson et al. 2010). Disorders of menstruation, which accompany puberty and reflect anovular ovarian cycles (Peacock et al. 2010), can be understood as years of education to fine-tune the TLR signaling cascade. Collectively, we assume that anovulation failures correspond with inadequate activation or inhibition of the inflammatory Myd88-dependent TLR signaling cascade being impaired at different levels of molecule activation by co-regulating pathways. Inhibition of the repair signalling then leads to luteinized cysts, which is the outcome of an insufficient connective tissue replacement of the former antrum and a frequent event in women of reproductive age. In the case of overactivation of the TRIF-dependent pathway, which causes the activation of IRF genes, the so-called IFN mRNA signature should be noted. It refers to inadequate clearance of apoptotic bodies shown, for example, in primary Sjögren's syndrome and systemic sclerosis (Meyer 2009). This connexion could also be correct for the many apoptotic bodies in the luteolytic CL of golden hamsters (Fig. 2.1e) and in regressing antral follicles (Fig. 3.2e, f).

The enormous self-healing potential of the ovary is amazing. It happens in spite of surgical interventions such as wedge resection or ovarian drilling. The strategy has minor effects on the follicle reserve; it is effective in women with PCOS, who have shown resistance to pharmacologically-induced ovulations (Api 2009). How ovarian surgery normalizes endocrine parameters in PCOS women is incompletely understood. Changes seem to be governed by the ovary itself and thus precede the restored feedback to the hypothalamus and pituitary gland (Hendriks et al. 2007). In view of this concept, we assume that INIM function is inhibited in polycystic ovaries and that tissue damage by ovarian drilling repairs blocked INIM interactions. Innate immunity does not stand alone as manager of ovarian functions, but it cooperates with adaptive immunity through DCs (Iwasaki and Medzhitov 2010; Peng et al. 2007; Turvey and Broide 2010). The statement also proves to be correct for the ovary, because T cells are reduced in number in the thecal cell layer of polycystic ovaries (Wu et al. 2007), and because an altered T-cell profile is reported for the follicular fluid of patients with idiopathic infertility (Lukassen et al. 2003). Furthermore, autoimmune damage is known to be responsible of premature ovarian failure diagnosed in women under the age of 40 years. The failure is associated with the alteration of T-cell subsets and T-cell-mediated cell injury (Vujovic 2009). The overall number of T cells is low in the preovulatory follicle and steadily increasing at the time of morphological luteolysis (Brännström et al. 1994b; Best et al. 1996; Bauer et al. 2001). The reports point to a more intense convertion between innate and adaptive immunity at times of luteolysis than of follicle rupture. It is noteworthy that impaired luteolysis never causes a tumor-like body. Another immune-mediated failure appears to be the luteal phase deficiency associated with failed or delayed implantation, infertility and early pregnancy loss (Erlebacher et al. 2004). In detail, when in a murine model the cell-mediated/adaptive response has been blocked by ligating CD40 (TNF receptor superfamily member and critical for adaptive immunity response), INIM is heavily activated, progesterone synthesis impaired and resistance to prolactin stimulation seen. These findings point to the interaction of INIM with the endocrine system. Finally, after menopause and absence of folliculogenesis, the original mission of INIM fades away, and mast cells completely disappear in the medulla and the intersitital cortical tissue (Heider et al. 2001). INIM might become involved in the progression of benign or malignant ovarian tumors. The possibility relates to the immunolocalization of multiple TLRs in the surface epithelium of human ovaries, in benign and malignant ovarian tumors (Zhou et al. 2009) and to prostate cancer risk associated with gene sequence variants in TLR4 as well as in TLR clusters (Girling and Hedger 2007). In the long run, novel strategies for the therapy of ovarian disorders will cross-react with the field of autoimmune diseases, allergy, transplantation and tumor biology.

Chapter 8
Concluding Summary and Remarks

For decades, it has been a black box to understand how tissue homeostasis is controlled in the ovary. Processes of proliferation such as growth and maturation of follicles and CL formation must be well balanced with events of involution such as follicular atresia and luteolysis. "Innate immunity" seems to be the answer that brings light into the complex biology of the ovary. The paradox seems that INIM exerts "non-immune" functions such as tissue damage (follicle rupture), repair (CL development) and tissue homeostasis (follicular atresia, luteolysis). The paradox is disentangled under the surmise that INIM reacts against danger signals either from outside of the body or from inside. One is excited to learn that evolution has tailored the capabilities of INIM to the demands of the ovary. It creates non lympoid effector cells with the potential of DC, which is the CK^+ cell. It is ingeniously hidden in growing follicles and in the CL of pregnancy, well seen in primordial follicles, in preovulatory follicles, and in CL of development (Figs. 4.2a–d and 4.3a–d). The CK expression seems to be associated with oxidative stress, which also explains the occurrence of CK^+ cells in regressing antral follicles (Fig. 4.2e–h). The majority represent CK^+ steroidogenic cells, which belong to granulosa cells and to luteal cells. The steroidogenic CK^+ cells seem to switch off CK expression and populate the CL of secretion and of pregnancy as granulosa-like CK^- cells, being type 5 in culture. The minority of CK^+ cells correlate with a microvascular type (type 1) that is hard to discern in the thecal cell layer of a preovulatory follicle (Fig. 4.3c) and clearly seen as part of the microvascular bed in the CL of secretion (Fig. 4.3d–f). The microvascular CK^+ cell can be isolated from the CL of development, secretion, and of regression (Fig. 5.12). The steroidogenic cell type is suggested to have originated from the surface epithelium of the genital ridge early in fetal life. The microvascular type comes from somite-derived angioblasts, which deliver endothelial precursor cells for migration into developing organ systems (Zovein et al. 2008).

The steroidogenic and microvascular subsets of CK^+ cells are promising candidates as DCs for the following reasons. The CK^+ granulosa cells regulate TLR4 and the co-regulatory CD14 receptor under oxLDL application (Serke et al. 2009, 2010). Type 5 cell as a postulated conversion from steroidogenic luteal CK^+ cells develops delicate finger-like processes (Fig. 5.11e, f), positively responds to INF-γ treatment with cell proliferation (Fig. 5.17), and an increase in N-cadherin contact proteins (Fig. 5.15i, j), and depicts the dramatic upregulation of MHC II peptides (Table 5.1). The microvascular CK^+ type 1 cells firmly increase tight junctions, adherens junctions and desmosomes under IFN-γ application (Fig. 5.15a–f). The intercellular contacts could be responsible for a closed microenvironment to handle danger sensing through the single cilium (Fig. 5.7b–e). Because vascular CK^+ cells are ubiquitous, they might represent specialized immune guardians and thus be a general principle of the circulation system.

We are at the beginning of analyzing the immunological dynamics of CK^+ cell subsets and of understanding their role in INIM function. The cross-talk between transduction pathways downstream of TLR receptors could vary for different groups of PRRs. The renaissance is anticipated for superovulated animal models, which will allow the studying of the time sequence of immunoregulatory genes, proteins and cytokine profiles. The entire complex of immunoresponses is far from being predicted for the ovulatory process, supposedly triggered by the kick of oxidative stress and conveying the utmost complex immune scenario of acute inflammation. Luteolysis, which compares with a chronic inflammation, should be studied under the aspect of INIM force. A possible autoimmune reaction has been forwarded as a mechanism of luteolysis more than two decades ago (Murdoch and Steadman 1991) and might be relevant in rapid luteolysis. It is connected with intraluteal hemorrage, vessel rupture and segregation of luteal cell complexes into the microvascular bed of the white-footed mouse, *Peromyscus leucopus* (Spanel-Borowski et al. 1983a). Interdisciplinary transfer of knowledge will help to understand immune mechanisms of sterile inflammations in other organs and diseases (Medzhitov 2008, 2010a).

The big adventure has begun to decipher the versatile machinery of INIM force in the ovary including the interaction with adaptive immunity and with the endocrine system. Unexplained aspects of INIM will be witnessed in future (Medzhitov 2010b). A great deal of work lies in the hands of the next generations to unravel the complexities of INIM in the biology and pathology of the ovary.

References

Adashi EY (1990) Do cytokines play a role in the regulation of ovarian function? Prog Neuroendocrinimmunol 3:11–17

Adashi EY (1994) Endocrinology of the ovary. Hum Reprod 9:815–827

Adashi EY, Leung PCK (eds) (1993) The ovary. Raven, New York

Agarwal A, Gupta S, Sharma R (2005) Oxidative stress and its implications in female infertility – a clinician's perspective. Reprod Biomed Online 11:641–650

Agarwal A, Saleh RA, Bedaiwy MA (2003) Role of reactive oxygen species in the pathophysiology of human reproduction. Fertil Steril 79:829–843

Antczak M, van Blerkom J (2000) The vascular character of ovarian follicular granulosa cells: phenotypic and functional evidence for an endothelial-like cell population. Hum Reprod 15:2306–2318

Api M (2009) Is ovarian reserve diminished after laparoscopic ovarian drilling? Gynecol Endocrinol 25:159–165

Arnold JN, Wallis R, Willis AC, Harvey DJ, Royle L, Dwek RA, Rudd PM, Sim RB (2006) Interaction of mannan binding lectin with alpha2 macroglobulin via exposed oligomannose glycans: a conserved feature of the thiol ester protein family? J Biol Chem 281:6955–6963

Auersperg N, Wong AS, Choi KC, Kang SK, Leung PC (2001) Ovarian surface epithelium: biology, endocrinology, and pathology. Endocr Rev 22:255–288

Aust G, Brylla E, Lehmann I, Kiessling S, Spitzer KL (1999) Different cytokine, adhesion molecule and prostaglandin receptor (PG-R) expression by cytokeratin 18 negative (CK−) and positive (CK+) endothelial cells (EC). Basic Res Cardiol 94:406

Aust G, Simchen C, Heider U, Hmeidan FA, Blumenauer V, Spanel-Borowski K (2000) Eosinophils in the human corpus luteum: the role of RANTES and eotaxin in eosinophil attraction into periovulatory structures. Mol Hum Reprod 6:1085–1091

Banchereau J, Steinman RM (1998) Dendritic cells and the control of immunity. Nature 392: 245–252

Bauer M, Reibiger I, Spanel-Borowski K (2001) Leucocyte proliferation in the bovine corpus luteum. Reproduction 121:297–305

Bausenwein J, Serke H, Eberle K, Hirrlinger J, Jogschies P, Hmeidan FA, Blumenauer V, Spanel-Borowski K (2010) Elevated levels of oxidized low-density lipoprotein and of catalase activity in follicular fluid of obese women. Mol Hum Reprod 16:117–124

Ben-Ze'ev A, Amsterdam A (1989) Regulation of cytoskeletal protein organization and expression in human granulosa. Endocrinology 124:1033–1041

Berisha B, Schams D (2005) Ovarian function in ruminants. Domest Anim Endocrinol 29: 305–317

Best CL, Pudney J, Welch WR, Burger N, Hill JA (1996) Localization and characterization of white blood cell populations within the the human ovary throughout the menstrual cycle and menopause. Hum Reprod 11:790–797

Bianchi ME (2007) DAMPs, PAMPs and alarmins: all we need to know about danger. J Leukoc Biol 81:1–5

Billiau A, Matthys P (2009) Interferon-gamma: a historical perspective. Cyst Growth Factor Rev 20:97–113

Blanchard C, Rothenberg ME (2009) Biology of the eosinophil. Adv Immunol 101:81–121

Blumenthal A, Ehlers S, Lauber J, Buer J, Lange C, Goldmann T, Heine H, Brandt E, Reiling N (2006) The Wingless homolog WNT5A and its receptor Frizzled-5 regulate inflammatory responses of human mononuclear cells induced by microbial stimulation. Blood 108:965–973

Bottazzi B, Bastone A, Doni A, Garlanda C, Valentino S, Deban L, Maina V, Cotena A, Moalli F, Vago L, Salustri A, Romani L, Mantovani A (2006) The long pentraxin PTX3 as a link among innate immunity, inflammation, and female fertility. J Leukoc Biol 79:909–912

Boyer A, Lapointe E, Zheng X, Cowan RG, Li H, Quirk SM, Demayo FJ, Richards JS, Boerboom D (2010) WNT4 is required for normal ovarian follicle development and female fertility. FASEB J [Epub ahead of print]

Brännström M, Enskog A (2002) Leukocyte networks and ovulation. J Reprod Immunol 57: 47–60

Brännström M, Mayrhofer G, Robertson SA (1993) Localization of leukocyte subsets in the rat ovary during the periovulatory period. Biol Reprod 48:277–286

Brännström M, Norman RJ, Seamark RF, Robertson SA (1994a) Rat ovary produces cytokines during ovulation. Biol Reprod 50:88–94

Brännström M, Pascoe V, Norman RJ, McClure N (1994b) Localization of leukocyte subsets in the follicle wall and in the corpus luteum. Fertil Steril 61:488–495

Brylla E, Aust G, Geyer M, Uckermann O, Löffler S, Spanel-Borowski K (2005) Coexpression of preprotachykinin A and B transcripts in the bovine corpus luteum and evidence for functional neurokinin receptor activity in luteal endothelial cells and ovarian macrophages. Regul Pept 125:125–133

Chan CW, Housseau F (2008) The 'kiss of death' by dendritic cells to cancer cells. Cell Death Differ 15:58–69

Chao W (2009) Toll-like receptor signaling: a critical modulator of cell survival and ischemic injury in the heart. Am J Physiol Heart Circ Physiol 296:H1–H12

Chen CJ, Kono H, Golenbock D, Reed G, Akira S, Rock KL (2007) Identification of a key pathway required for the sterile inflammatory response triggered by dying cells. Nat Med 13:851–856

Chorro L, Sarde A, Li M, Woollard KJ, Chambon P, Malissen B, Kissenpfennig A, Barbaroux JB, Groves R, Geissmann F (2009) Langerhans cell (LC) proliferation mediates neonatal development, homeostasis, and inflammation-associated expansion of the epidermal LC network. J Exp Med 206:3089–3100

Craig J, Orisaka M, Wang H, Orisaka S, Thompson W, Zhu C, Kotsuji F, Tsang BK (2007) Gonadotropin and intra-ovarian signals regulating follicle development and atresia: the delicate balance between life and death. Front Biosci 12:3628–3639

Criado-Garcia O, Fernaud-Espinosa I, Bovolenta P, Sainz L, de la Cuesta R, Rodriguez de Córdoba S (1999) Expression of the beta-chain of the complement regulator C4b-binding protein in human ovary. Eur J Cell Biol 78:657–664

Czernobilsky B, Moll R, Levy R, Franke WW (1985) Co-expression of cytokeratin and vimentin filaments in mesothelial, granulosa and rete ovarii cells of the human ovary. Eur J Cell Biol 37:175–190

Dandapat A, Hu C, Sun L, Mehta JL (2007) Small concentrations of oxLDL induce capillary tube formation from endothelial cells via LOX-1-dependent redox-sensitive pathway. Arterioscler Thromb Vasc Biol 27:2435–2442

Davis JS, Rueda BR, Spanel-Borowski K (2003) Microvascular endothelial cells of the corpus luteum. Reprod Biol Endocrinol 1:89

Davis EE, Brueckner M, Katsanis N (2006) The emerging complexity of the vertebrate cilium: new functional roles for an ancient organelle. Dev Cell 11:9–19

Debeljuk L (2006) Tachykinins and ovarian function in mammals. Peptides 27:736–742

Del Canto F, Sierralta W, Kohen P, Munoz A, Strauss JF 3rd, Devoto L (2007) Features of natural and gonadotropin-releasing hormone antagonist-induced corpus. J Clin Endocrinol Metab 92:4436–4443
Derouet-Humbert E, Roemer K, Bureik M (2005) Adrenodoxin (Adx) and CYP11A1 (P450scc) induce apoptosis by the generation of reactive oxygen species in mitochondria. Biol Chem 386:453–461
Devoto L, Fuentes A, Kohen P, Cespedes P, Palomino A, Pommer R, Munoz A, Strauss JF 3rd (2009) The human corpus luteum: life cycle and function in natural cycles. Fertil Steril 92:1067–1079
Dieterlen-Lièvre F, Pouget C, Bollérot K, Jaffredo T (2006) Are intra-aortic hemopoietic cells derived from endothelial cells during ontogeny? Trends Cardiovasc Med 16:128–139
DiScipio RG, Schraufstatter IU (2007) The role of the complement anaphylatoxins in the recruitment of eosinophils. Int Immunopharmacol 7:1909–1923
Doan N, Gettins PG (2007) Human alpha2-macroglobulin is composed of multiple domains, as predicted by homology with complement component C3. Biochem J 407:23–30
Duerrschmidt N, Zabirnyk O, Nowicki M, Ricken A, Hmeidan FA, Blumenauer V, Borlak J, Spanel-Borowski K (2006) Lectin-like oxidized low-density lipoprotein receptor-1-mediated autophagy in human granulosa cells as an alternative of programmed cell death. Endocrinology 147:3851–3860
Dunzendorfer S, Wiedermann CJ (2001) Neuropeptides and the immune system: focus on dendritic cells. Crit Rev Immunol 21:523–557
Endo Y, Takahashi M, Fujita T (2006) Lectin complement system and pattern recognition. Immunobiology 211:283–293
Erlebacher A, Zhang D, Parlow AF, Glimcher LH (2004) Ovarian insufficiency and early pregnancy loss induced by activation of the innate immune system. J Clin Invest 114:39–48
Espey LL (1994) Current status of the hypothesis that mammalian ovulation is comparable to an inflammatory reaction. Biol Reprod 50:233–238
Espey LL (2006) Comprehensive analysis of ovarian gene expression during ovulation using differential display. Methods Mol Biol 317:219–241
Fenyves AM, Behrens J, Spanel-Borowski K (1993) Cultured microvascular endothelial cells (MVEC) differ in cytoskeleton, expression of cadherins and fibronectin matrix. A study under the influence of interferon-gamma. J Cell Sci 106:879–890
Fenyves AM, Saxer M, Spanel-Borowski K (1994) Bovine microvascular endothelial cells of separate morphology differ in growth and response to the action of interferon-gamma. Experientia 50:99–104
Ferguson TA, Green DR, Griffith TS (2002) Cell death and immune privilege. Int Rev Immunol 21:153–172
Foell D, Wittkowski H, Vogl T, Roth J (2007) S100 proteins expressed in phagocytes: a novel group of damage-associated molecular pattern molecules. J Leukoc Biol 81:28–37
Fraser HM, Duncan WC (2005) Vascular morphogenesis in the primate ovary. Angiogenesis 8:101–116
Fraser HM, Wulff C (2001) Angiogenesis in the primate ovary. Reprod Fertil Dev 13:557–566
Fraser HM, Wulff C (2003) Angiogenesis in the corpus luteum. Reprod Biol Endocrinol 1:88
Friesel R, Komoriya A, Maciag T (1987) Inhibition of endothelial cell proliferation by gamma-interferon. J Cell Biol 104:689–696
Gaytn M, Morales C, Sanchez-Criado JE, Gaytn F (2008) Immunolocalization of beclin 1, a bcl-2-binding, autophagy-related protein, in the human ovary: possible relation to life span of corpus luteum. Cell Tissue Res 331:509–517
Geissmann F, Manz MG, Jung S, Sieweke MH, Merad M, Ley K (2010) Development of monocytes, macrophages, and dendritic cells. Science 327:656–661
Ghai R, Waters P, Roumenina LT, Gadjeva M, Kojouharova MS, Reid KB, Sim RB, Kishore U (2007) C1q and its growing family. Immunobiology 212:253–266

Girling JE, Hedger MP (2007) Toll-like receptors in the gonads and reproductive tract: emerging roles in reproductive physiology and pathology. Immunol Cell Biol 85:481–489

Gordon MD, Dionne MS, Schneider D, Nusse R (2005) WntD is a feedback inhibitor of Dorsal/NF-kappaB in *Drosophila* development and immunity. Nature 437:746–749

Gougeon A (1993) Dynamics of human follicular growth: a morphologic perspective. In: Adashi EY, Leung PCK (eds) The ovary. Raven, New York, pp 21–39

Grissell TV, Chang AB, Gibson PG (2007) Reduced toll-like receptor 4 and substance P gene expression is associated with airway bacterial colonization in children. Pediatr Pulmonol 42:380–385

Hajishengallis G, Lambris JD (2010) Crosstalk pathways between Toll-like receptors and the complement system. Trends Immunol 31:154–163

Hanukoglu I (2006) Antioxidant protective mechanisms against reactive oxygen species (ROS) generated by mitochondrial P450 systems in steroidogenic cells. Drug Metab Rev 38:171–196

Hartvigsen K, Chou MY, Hansen LF, Shaw PX, Tsimikas S, Binder CJ, Witztum JL (2009) The role of innate immunity in atherogenesis. J Lipid Res 50(Suppl):S388–S393

Hawlisch H, Köhl J (2006) Complement and Toll-like receptors: key regulators of adaptive immune responses. Mol Immunol 43:13–21

He J, Xiao Z, Chen X, Chen M, Fang L, Yang M, Lv Q, Li Y, Li G, Hu J, Xie X (2010) The expression of functional toll-like receptor 4 is associated with proliferation and maintenance of stem cell phenotype in endothelial progenitor cells (EPCs). J Cell Biochem 2010 [Epub ahead of print]

Heider U, Pedal I, Spanel-Borowski K (2001) Increase in nerve fibers and loss of mast cells in polycystic and postmenopausal ovaries. Fertil Steril 75:1141–1147

Hendriks ML, Ket JC, Hompes PG, Homburg R, Lambalk CB (2007) Why does ovarian surgery in PCOS help? Insight into the endocrine implications of ovarian surgery for ovulation induction in polycystic ovary syndrome. Hum Reprod Update 13:249–264

Herath S, Williams EJ, Lilly ST, Gilbert RO, Dobson H, Bryant CE, Sheldon IM (2007) Ovarian follicular cells have innate immune capabilities that modulate their endocrine function. Reproduction 134:683–693

Hernandez-Gonzalez I, Gonzalez-Robayna I, Shimada M, Wayne CM, Ochsner SA, White L, Richards JS (2006) Gene expression profiles of cumulus cell oocyte complexes during ovulation reveal cumulus cells express neuronal and immune-related genes: does this expand their role in the ovulation process? Mol Endocrinol 20:1300–1321

Herrman G, Missfelder H, Spanel-Borowski K (1996) Lectin binding patterns in two cultured endothelial cell types derived from bovine corpus luteum. Histochem Cell Biol 105:129–137

Horne AW, Stock SJ, King AE (2008) Innate immunity and disorders of the female reproductive tract. Reproduction 135:739–749

Hoshino K, Kaisho T (2008) Nucleic acid sensing Toll-like receptors in dendritic cells. Curr Opin Immunol 20:408–413

Hussein MR (2005) Apoptosis in the ovary: molecular mechanisms. Hum Reprod Update 11:162–177

Iijima N, Thompson JM, Iwasaki A (2008) Dendritic cells and macrophages in the genitourinary tract. Mucosal Immunol 1:451–459

Ireland JL, Jimenez-Krassel F, Winn ME, Burns DS, Ireland JJ (2004) Evidence for autocrine or paracrine roles of alpha2-macroglobulin in regulation. Endocrinology 145:2784–2794

Iwasaki A, Medzhitov R (2010) Regulation of adaptive immunity by the innate immune system. Science 327:291–295

Jahn L, Fouquet B, Rohe K, Franke WW (1987) Cytokeratins in certain endothelial and smooth muscle cells of two taxonomically distant vertebrate species, *Xenopus laevis* and man. Differentiation 36:234–254

Jakob W, Jentzsch KD, Mauersberger B, Oehme P (1977) Demonstration of angiogenesis-activity in the corpus luteum of cattle. Exp Pathol (Jena) 13:231–236

Jarkovska K, Martinkova J, Liskova L, Halada P, Moos J, Rezabek K, Gadher SJ, Kovarova H (2010) Proteome mining of human follicular fluid reveals a crucial role of complement. J Proteome Res 9:1289–1301

Kahnberg A, Enskog A, Brännström M, Lundin K, Bergh C (2009) Prediction of ovarian hyperstimulation syndrome in women undergoing in vitro fertilization. Acta Obstet Gynecol Scand 88:1373–1381

Kalesnikoff J, Galli SJ (2008) New developments in mast cell biology. Nat Immunol 9: 1215–1223

Keller M, Ruegg A, Werner S, Beer HD (2008) Active caspase-1 is a regulator of unconventional protein secretion. Cell 132:818–831

Khan-Dawood FS, Yusoff DM, Tabibzadeh S (1996) Immunohistochemical analysis of the microanatomy of primate ovary. Biol Reprod 54:734–742

Kim J, Bagchi IC, Bagchi MK (2009) Signaling by hypoxia-inducible factors is critical for ovulation in mice. Endocrinology 150:3392–3400

Klassert TE, Pinto F, Hernandez M, Candenas ML, Hernandez MC, Abreu J, Almeida TA (2008) Differential expression of neurokinin B and hemokinin-1 in human immune cells. J Neuroimmunol 196:27–34

Klune JR, Dhupar R, Cardinal J, Billiar TR, Tsung A (2008) HMGB1: endogenous danger signaling. Mol Med 14:476–484

Knapp AC, Franke WW (1989) Spontaneous losses of control of cytokeratin gene expression in transformed, non-epithelial human cells occurring at different levels of regulation. Cell 59:67–79

Koch D, Sakurai M, Hummitzsch K, Hermsdorf T, Erdmann S, Schwalbe S, Stolzenburg JU, Spanel-Borowski K, Ricken AM (2009) KIT variants in bovine ovarian cells and corpus luteum. Growth Factors 27:100–113

Kohchi C, Inagawa H, Nishizawa T, Soma G (2009) ROS and innate immunity. Anticancer Res 29:817–821

Köhl J (2006a) Self, non-self, and danger: a complementary view. Adv Exp Med Biol 586:71–94

Köhl J (2006b) The role of complement in danger sensing and transmission. Immunol Res 34:157–176

Krohn PL (1977) Transplantation of the ovary. In: Zuckerman S, Weir B (eds) The ovary, vol 2. Academic, New York, pp 101–127

Kuman RJ, Shih I-M (2010) The origin and pathogenesis of epithelial ovarian cancer: a proposed unifying theory. Am J Surg Pathol 34:433–443

Kumar H, Kawai T, Akira S (2009) Toll-like receptors and innate immunity. Biochem Biophys Res Commun 388:621–625

Lambrecht BN (2001) Immunologists getting nervous: neuropeptides, dendritic cells and T cell. Respir Res 2:133–138

Lehmann I, Brylla E, Sittig D, Spanel-Borowski K, Aust G (2000) Microvascular endothelial cells differ in their basal and tumour necrosis factor-alpha-regulated expression of adhesion molecules and cytokines. J Vasc Res 37:408–416

Ley K, Gaehtgens P, Spanel-Borowski K (1992) Differential adhesion of granulocytes to five distinct phenotypes of cultured microvascular endothelial cells. Microvasc Res 43:119–133

Lindner I, Hemdan NY, Buchold M, Huse K, Bigl M, Oerlecke I, Ricken A, Gaunitz F, Sack U, Naumann A, Hollborn M, Thal D, Gebhardt R, Birkenmeier G (2010) Alpha2-macroglobulin inhibits the malignant properties of astrocytoma cells by impeding beta-catenin signaling. Cancer Res 70:277–287

Liu Z, Shimada M, Richards JS (2008) The involvement of the Toll-like receptor family in ovulation. J Assist Reprod Genet 25:223–228

Löffler S, Horn LC, Weber W, Spanel-Borowski K (2000) The transient disappearance of cytokeratin in human fetal and adult ovaries. Anat Embryol (Berl) 201:207–215

Löffler S, Schulz A, Brylla E, Nieber K, Spanel-Borowski K (2004a) Transcripts of neurokinin B and neurokinin 3 receptor in superovulated rat ovaries and increased number of corpora lutea as a non-specific effect of intraperitoneal agonist application. Regul Pept 122:131–137

Löffler S, Schulz A, Hunt SP, Spanel-Borowski K (2004b) Increased formation of corpora lutea in neurokinin 1-receptor deficient mice. Mol Reprod Dev 68:408–414

Löseke A, Spanel-Borowski K (1996) Simple or repeated induction of superovulation: a study on ovulation rates and microvessel corrosion casts in ovaries of golden hamsters. Ann Anat 178:5–14

Lukassen HG, van der Meer A, van Lierop MJ, Lindeman EJ, Joosten I, Braat DD (2003) The proportion of follicular fluid CD16+CD56DIM NK cells is increased in IVF patients with idiopathic infertility. J Reprod Immunol 60:71–84

Matsuda-Minehata F, Inoue N, Goto Y, Manabe N (2006) The regulation of ovarian granulosa cell death by pro- and anti-apoptotic. J Reprod Dev 52:695–705

Matsushita M (2009) Ficolins: complement-activating lectins involved in innate immunity. J Innate Immun 2:24–32

Matzinger P (2002) The danger model: a renewed sense of self. Science 296:301–305

Matzinger P (2007) Friendly and dangerous signals: is the tissue in control? Nat Immunol 8:11–13

Mayer G (2009) Innate (non-specific) immunity. Immunology – chapter one. In: University of South Carolina (ed) Microbiology and immunology. Online textbook, pp 1–10

Mayerhofer A, Spanel-Borowski K, Watkins S, Gratzl M (1992) Cultured microvascular endothelial cells derived from the bovine corpus luteum possess NCAM-140. Exp Cell Res 201:545–548

McAuslan BR, Hannan GN, Reilly W (1982) Signals causing change in morphological phenotype, growth mode, and gene expression of vascular endothelial cells. J Cell Physiol 112:96–106

Medzhitov R (2008) Origin and physiological roles of inflammation. Nature 454:428–435

Medzhitov R (2010a) Inflammation 2010: new adventures of an old flame. Cell 140:771–776

Medzhitov R (2010b) Innate immunity: quo vadis? Nat Immunol 11:551–553

Mehta JL, Chen J, Hermonat PL, Romeo F, Novelli G (2006) Lectin-like, oxidized low-density lipoprotein receptor-1 (LOX-1): a critical player in the development of atherosclerosis and related disorders. Cardiovasc Res 69:36–45

Mei J, Chen B, Yue H, Gui JF (2008) Identification of a C1q family member associated with cortical granules and follicular cell apoptosis in *Carassius auratus gibelio*. Mol Cell Endocrinol 289:67–76

Mellman I, Steinman RM (2001) Dendritic cells: specialized and regulated antigen processing machines. Cell 106:255–258

Merad M, Manz MG (2009) Dendritic cell homeostasis. Blood 113:3418–3427

Merkwitz C, Ricken AM, Lösche A, Sakurai M, Spanel-Borowski K (2010) Progenitor cells harvested from bovine follicles become endothelial cells. Differentiation 79:203–210

Meyer O (2009) Interferons and autoimmune disorders. Joint Bone Spine 76:464–473

Michelsen KS, Wong MH, Shah PK, Zhang W, Yano J, Doherty TM, Akira S, Rajavashisth TB, Arditi M (2004) Lack of Toll-like receptor 4 or myeloid differentiation factor 88 reduces atherosclerosis and alters plaque phenotype in mice deficient in apolipoprotein E. Proc Natl Acad Sci U S A 101:10679–10684

Miller YI, Chang MK, Binder CJ, Shaw PX, Witztum JL (2003a) Oxidized low density lipoprotein and innate immune receptors. Curr Opin Lipidol 14:437–445

Miller YI, Viriyakosol S, Binder CJ, Feramisco JR, Kirkland TN, Witztum JL (2003b) Minimally modified LDL binds to CD14, induces macrophage spreading via TLR4/MD-2, and inhibits phagocytosis of apoptotic cells. J Biol Chem 278:1561–1568

Mineau-Hanschke R, Patton WF, Hechtman HB, Shepro D (1993) Immunolocalization of cytokeratin 19 in bovine and human pulmonary microvascular endothelial cells in situ. Comp Biochem Physiol Comp Physiol 104:313–319

Munitz A, Levi-Schaffer F (2004) Eosinophils: 'new' roles for 'old' cells. Allergy 59:268–275
Murdoch WJ, Steadman LE (1991) Investigations concerning the relationship of ovarian eosinophilia to ovulation. Am J Reprod Immunol 25:81–87
Nicosia SV, Johnson JH (1984) Surface morphology of ovarian mesothelium (surface epithelium) and of other pelvic and extrapelvic mesothelial sites in the rabbit. Int J Gynecol Pathol 3:249–260
Niederkorn JY (2006) See no evil, hear no evil, do no evil: the lessons of immune privilege. Nat Immunol 7:354–359
Nishimura R, Okuda K (2010) Hypoxia is important for establishing vascularisation during corpus luteum formation in cattle. J Reprod Dev 56:110–116
Nissim Ben Efraim AH, Levi-Schaffer F (2008) Tissue remodeling and angiogenesis in asthma: the role of the eosinophil. Ther Adv Respir Dis 2:163–171
Niswender GD, Juengel JL, Silva PJ, Rollyson MK, McIntush EW (2000) Mechanisms controlling the function and life span of the corpus luteum. Physiol Rev 80:1–29
O'Connor TM, O'Connell J, O'Brien DI, Goode T, Bredin C, Shanahan F (2004) The role of substance P in inflammatory disease. J Cell Physiol 201:167–180
Oglesby TJ, Longwith JE, Huettner PC (1996) Human complement regulator expression by the normal female reproductive tract. Anat Rec 246:78–86
Oktem O, Oktay K (2008) The ovary: anatomy and function throughout human life. Ann NY Acad Sci 1127:1–9
O'Neill LA (2008) The interleukin-1 receptor/Toll-like receptor superfamily: 10 years of progress. Immunol Rev 226:10–18
O'Neill LA, Bowie AG (2007) The family of five: TIR-domain-containing adaptors in Toll-like receptor. Nat Rev Immunol 7:353–364
Page NM, Bell NJ, Gardiner SM, Manyonda IT, Brayley KJ, Strange PG, Lowry PJ (2003) Characterization of the endokinins: human tachykinins with cardiovascular activity. Proc Natl Acad Sci U S A 100:6245–6250
Pan J, Auersperg N (1998) Spatiotemporal changes in cytokeratin expression in the neonatal rat ovary. Biochem Cell Biol 76:27–35
Paterson HM, Murphy TJ, Purcell EJ, Shelley O, Kriynovich SJ, Lien E, Mannick JA, Lederer JA (2003) Injury primes the innate immune system for enhanced Toll-like receptor reactivity. J Immunol 171:1473–1483
Patton WF, Yoon MU, Alexander JS, Chung-Welch N, Hechtman HB, Shepro D (1990) Expression of simple epithelial cytokeratins in bovine pulmonary microvascular. J Cell Physiol 143:140–149
Peacock A, Alvi NS, Mushtaq T (2010) Period problems: disorders of menstruation in adolescents. Arch Dis Child [Epub ahead of print]
Peng Y, Martin DA, Kenkel J, Zhang K, Ogden CA, Elkon KB (2007) Innate and adaptive immune response to apoptotic cells. J Autoimmun 29:303–309
Pereira CP, Bachli EB, Schoedon G (2009) The wnt pathway: a macrophage effector molecule that triggers inflammation. Curr Atheroscler Rep 11:236–242
Peters KG, Kontos CD, Lin PC, Wong AL, Rao P, Huang L, Dewhirst MW, Sankar S (2004) Functional significance of Tie2 signaling in the adult vasculature. Recent Prog Horm Res 59:51–71
Pouget C, Gautier R, Teillet MA, Jaffredo T (2006) Somite-derived cells replace ventral aortic hemangioblasts and provide aortic smooth muscle cells of the trunk. Development 133:1013–1022
Qublan H, Amarin Z, Nawasreh M, Diab F, Malkawi S, Al-Ahmad N, Balawneh M (2006) Luteinized unruptured follicle syndrome: incidence and recurrence rate in infertile women with unexplained infertility undergoing intrauterine insemination. Hum Reprod 21:2110–2113
Rachon D, Teede H (2010) Ovarian function and obesity-interrelationship, impact on women's reproductive lifespan and treatment options. Mol Cell Endocrinol 316:172–179

Rajah R, Glaser EM, Hirshfield AN (1992) The changing architecture of the neonatal rat ovary during histogenesis. Dev Dyn 194:177–192

Reed-Geaghan EG, Savage JC, Hise AG, Landreth GE (2009) CD14 and toll-like receptors 2 and 4 are required for fibrillar A{beta}-stimulated microglial activation. J Neurosci 29:11982–11992

Reibiger I, Spanel-Borowski K (2000) Difference in localization of eosinophils and mast cells in the bovine ovary. J Reprod Fertil 118:243–249

Reibiger I, Aust G, Tscheudschilsuren G, Beyer R, Gebhardt C, Spanel-Borowski K (2001) The expression of substance P and its neurokinin-1 receptor mRNA in the bovine corpus luteum of early developmental stage. Neurosci Lett 299:49–52

Reich R, Miskin R, Tsafriri A (1985) Follicular plasminogen activator: involvement in ovulation. Endocrinology 116:516–521

Rescigno M, Urbano M, Valzasina B, Francolini M, Rotta G, Bonasio R, Granucci F, Kraehenbuhl JP, Ricciardi-Castagnoli P (2001) Dendritic cells express tight junction proteins and penetrate gut epithelial monolayers to sample bacteria. Nat Immunol 2:361–367

Reynolds LP, Grazul-Bilska AT, Redmer DA (2000) Angiogenesis in the corpus luteum. Endocrine 12:1–9

Richards JS, Russell DL, Ochsner S, Espey LL (2002) Ovulation: new dimensions and new regulators of the inflammatory-like response. Annu Rev Physiol 64:69–92

Richards JS, Liu Z, Shimada M (2008) Immune-like mechanisms in ovulation. Trends Endocrinol Metab 19:191–196

Ricken AM, Spanel-Borowski K (1996) Immunolocalization of neurophysin in cytokeratin-positive luteal cells of cows. Histochem Cell Biol 106:487–493

Ricken AM, Spanel-Borowski K, Saxer M, Huber PR (1995) Cytokeratin expression in bovine corpora lutea. Histochem Cell Biol 103:345–354

Ricken A, Rahner C, Landmann L, Spanel-Borowski S (1996) Bovine endothelial-like cells increase intercellular junctions under treatment with interferon-gamma. An in vitro study. Ann Anat 178:321–330

Rock KL, Latz E, Ontiveros F, Kono H (2010) The sterile inflammatory response. Annu Rev Immunol 28:321–342

Rodgers RJ, Irving-Rodgers HF (2010) Morphological classification of bovine ovarian follicles. Reproduction 139:309–318

Rohm F, Spanel-Borowski K, Eichler W, Aust G (2002) Correlation between expression of selectins and migration of eosinophils into the bovine ovary during the periovulatory period. Cell Tissue Res 309:313–322

Rojas D, Krishnan R (2010) IFN-gamma generates maturation-arrested dendritic cells that induce T cell hyporesponsiveness independent of Foxp3 (+) T-regulatory cell generation. Immunol Lett [Epub ahead of print]

Salvayre R, Auge N, Benoist H, Negre-Salvayre A (2002) Oxidized low-density lipoprotein-induced apoptosis. Biochim Biophys Acta 1585:213–221

Sandri S, Rodriguez D, Gomes E, Monteiro HP, Russo M, Campa A (2008) Is serum amyloid A an endogenous TLR4 agonist? J Leukoc Biol 83:1174–1180

Santini D, Ceccarelli C, Mazzoleni G, Pasquinelli G, Jasonni VM, Martinelli GN (1993) Demonstration of cytokeratin intermediate filaments in oocytes of the developing and adult human ovary. Histochemistry 99:311–319

Schams D, Berisha B (2004) Regulation of corpus luteum function in cattle – an overview. Reprod Domest Anim 39:241–251

Seimon TA, Obstfeld A, Moore KJ, Golenbock DT, Tabas I (2006) Combinatorial pattern recognition receptor signaling alters the balance of life and death in macrophages. Proc Natl Acad Sci U S A 103:19794–19799

Serke H, Vilser H, Nowicki M, Hmeidan FA, Blumenauer V, Hummitzsch K, Lösche A, Spanel-Borowski K (2009) Granulosa cell subtypes respond by autophagy or cell death

to oxLDL-dependent activation of the oxidized lipoprotein receptor 1 and toll-like 4 receptor. Autophagy 5:991–1003
Serke H, Bausenwein J, Hirrlinger J, Nowicki M, Vilser C, Jogschies P, Hmeidan FA, Blumenauer V, Spanel-Borowski K (2010) Granulosa cell subtypes vary in response to oxidized low-density lipoprotein as regards specific lipoprotein receptors and antioxidant enzyme activity. J Clin Endocrinol Metab 95:3480–3490
Severini C, Improta G, Falconieri-Erspamer G, Salvadori S, Erspamer V (2002) The tachykinin peptide family. Pharmacol Rev 54:285–322
Shimada M, Hernandez-Gonzalez I, Gonzalez-Robanya I, Richards JS (2006) Induced expression of pattern recognition receptors in cumulus oocyte complexes: novel evidence for innate immune-like functions during ovulation. Mol Endocrinol 20:3228–3239
Shimada M, Yanai Y, Okazaki T, Noma N, Kawashima I, Mori T, Richards JS (2008) Hyaluronan fragments generated by sperm-secreted hyaluronidase stimulate cytokine/chemokine production via the TLR2 and TLR4 pathway in cumulus cells of ovulated COCs, which may enhance fertilization. Development 135:2001–2011
Singla V, Reiter JF (2006) The primary cilium as the cell's antenna: signaling at a sensory organelle. Science 313:629–633
Souza DG, Soares AC, Pinho V, Torloni H, Reis LF, Teixeira MM, Dias AA (2002) Increased mortality and inflammation in tumour necrosis factor-stimulated gene-14 transgenic mice after ischemia and reperfusion injury. Am J Pathol 160:1755–1765
Spanel-Borowski K (1981) Morphological investigations on follicular atresia in canine ovaries. Cell Tissue Res 214:155–168
Spanel-Borowski K (1987) Fibrinolytic activity in ovaries of rats and golden hamsters after gonadotrophic stimulation and hypophysectomy. Anim Reprod Sci 14:301–307
Spanel-Borowski K (1989) Vascularization of ovaries from golden hamsters following implantation into the chick chorioallantoic membrane. Exp Cell Biol 57:219–227
Spanel-Borowski K (1991) Diversity of ultrastructure in different phenotypes of cultured microvessel endothelial cells isolated from bovine corpus luteum. Cell Tissue Res 266:37–49
Spanel-Borowski K, Aumüller G (1985) Light and ultrastructure of intra-ovarian oocyte release in infantile rats. Anat Embryol (Berl) 172:331–337
Spanel-Borowski K, Bein G (1993) Different microvascular endothelial cell phenotypes exhibit different class I and II antigens under interferon-gamma. In Vitro Cell Dev Biol Anim 29A:601–602
Spanel-Borowski K, Heiss C (1986) Luteolysis and thrombus formation in ovaries of immature superstimulated golden hamsters. Aust J Biol Sci 39:407–416
Spanel-Borowski K, Ricken AM (1997) Varying morphology of bovine granulosa cell cultures. In: Motta PM (ed) Microscopy of reproduction and development: a dynamic approach, Antonio Delfino, Rome, pp 91–100
Spanel-Borowski K, van der Bosch J (1990) Different phenotypes of cultured microvessel endothelial cells obtained from bovine corpus luteum. Study by light microscopy and by scanning electron microscopy (SEM). Cell Tissue Res 261:35–47
Spanel-Borowski K, Petterborg LJ, Reiter RJ (1982) Preantral intra-ovarian oocyte release in the white-footed mouse, Peromyscus leucopus. Cell Tissue Res 226:461–464
Spanel-Borowski K, Bartke A, Petterborg LJ, Reiter RJ (1983a) A possible mechanism of rapid luteolysis in white-footed mice, *Peromyscus leucopus*. J Morphol 176:225–233
Spanel-Borowski K, Vaughan LY, Johnson LY, Reiter RJ (1983b) Increase of intra-ovarian oocyte release in PMSG-primed immature rats and its inhibition by arginine vasotocin. Biomed Res 4:71–82
Spanel-Borowski K, Thor-Wiedemann S, Pilgrim C (1984) Cell proliferation in the dog (beagle) ovary during proestrus and early estrus. Acta Anat (Basel) 118:153–158
Spanel-Borowski K, Sohn G, Schlegel W (1986) Effects of locally applied enzyme inhibitors of the arachidonic acid cascade on follicle growth and intra-ovarian oocyte release in hyperstimulated rabbits. Arch Histol Jpn 49:565–577

Spanel-Borowski K, Amselgruber W, Sinowatz F (1987) Capilary sprouts in ovaries of immature superstimulated golden hamsters: a SEM study of microcorrosion casts. Anat Embryol 176:387–391

Spanel-Borowski K, Ricken AM, Kress A, Huber PR (1994a) Isolation of granulosa-like cells from the bovine secretory corpus luteum and their characterization in long-term culture. Anat Rec 239:269–279

Spanel-Borowski K, Ricken AM, Patton WF (1994b) Cytokeratin-positive and cytokeratin-negative cultured endothelial cells from bovine aorta and vena cava. Differentiation 57:225–234

Spanel-Borowski K, Sass K, Löffler S, Brylla E, Sakurai M, Ricken AM (2007) KIT receptor-positive cells in the bovine corpus luteum are primarily theca-derived small luteal cells. Reproduction 134:625–634

Standaert FE, Zamora CS, Chew BP (1991) Quantitative and qualitative changes in blood leukocytes in the porcine ovary. Am J Reprod Immunol 25:163–168

Steinman RM, Banchereau J (2007) Taking dendritic cells into medicine. Nature 449:419–426

Stewart CR, Stuart LM, Wilkinson K, van Gils JM, Deng J, Halle A, Rayner KJ, Boyer L, Zhong R, Frazier WA, Lacy-Hulbert A, Khoury JE, Golenbock DT, Moore KJ (2010) CD36 ligands promote sterile inflammation through assembly of a Toll-like receptor 4 and 6 heterodimer. Nat Immunol 11:155–161

Stocco C, Telleria C, Gibori G (2007) The molecular control of corpus luteum formation, function, and regression. Endocr Rev 28:117–149

Stolpen AH, Guinan EC, Fiers W, Pober JS (1986) Recombinant tumor necrosis factor and immune interferon act singly and in combination to reorganize human vascular endothelial cell monolayers. Am J Pathol 123:16–24

Stouffer RL, Xu F, Duffy DM (2007) Molecular control of ovulation and luteinization in the primate follicle. Front Biosci 12:297–307

Takeda K, Akira S (2005) Toll-like receptors in innate immunity. Int Immunol 17:1–14

Takeuchi O, Akira S (2010) Pattern recognition receptors and inflammation. Cell 140:805–820

Thomson RL, Buckley JD, Brinkworth GD (2010) Exercise for the treatment and management of overweight women with polycystic ovary syndrome: a review of the literature. Obes Rev [Epub ahead of print]

Tscheudschilsuren G, Aust G, Nieber K, Schilling N, Spanel-Borowski K (2002) Microvascular endothelial cells differ in basal and hypoxia-regulated expression of angiogenic factors and their receptors. Microvasc Res 63:243–251

Tsikolia N, Merkwitz C, Sass K, Sakurai M, Spanel-Borowski K, Ricken AM (2009) Characterization of bovine fetal Leydig cells by KIT expression. Histochem Cell Biol 132:623–632

Tsung A, Zheng N, Jeyabalan G, Izuishi K, Klune JR, Geller DA, Lotze MT, Lu L, Billiar TR (2007) Increasing numbers of hepatic dendritic cells promote HMGB1-mediated ischemia-reperfusion injury. J Leukoc Biol 81:119–128

Turvey SE, Broide DH (2010) Innate immunity. J Allergy Clin Immunol 125:S24–S32

Valeur HS, Valen G (2009) Innate immunity and myocardial adaptation to ischemia. Basic Res Cardiol 104:22–32

van den Hurk R, Dijkstra G, van Mill FN, Hulshof SC, van den Ingh TS (1995) Distribution of the intermediate filament proteins vimentin, keratin, and desmin. Mol Reprod Dev 41:459–467

van Lierop PPE, Samsom JN, Escher JC, Nieuwenhuis EE (2009) Role of the innate immune system in the pathogenesis of inflammatory bowel disease. J Pediatr Gastroenterol Nutr 48:142–151

Van Wenzel IL, Dharmarajan AM, Lavranos TC, Rodgers RJ (1999) Evidence for alternative pathways of granulosa cell death in healthy and slightly. Endocrinology 140:2602–2612

Vilser C, Hueller H, Nowicki M, Hmeidan FA, Blumenauer V, Spanel-Borowski K (2010) The variable expression of lectin-like oxidized low-density lipoprotein receptor (LOX-1) and signs of autophagy and apoptosis in freshly harvested human granulosa cells depend on gonadotropin dose, age, and body weight. Fertil Steril 93:2706–2715

Vujovic S (2009) Aetiology of premature ovarian failure. Menopause Int 15:72–75
Wasiuk A, de Vries VC, Hartmann K, Roers A, Noelle RJ (2009) Mast cells as regulators of adaptive immunity to tumours. Clin Exp Immunol 155:140–146
Wild RA, Carmina E, Diamanti-Kandarakis E, Dokras A, Escobar-Morreale HF, Futterweit W, Lobo R, Norman RJ, Talbott E, Dumesic DA (2010) Assessment of cardiovascular risk and prevention of cardiovascular disease in women with the polycystic ovary syndrome: a consensus statement by the Androgen Excess and Polycystic Ovary Syndrome (AE-PCOS) Society. J Clin Endocrinol Metab 95:2038–2049
Wolf KW, Spanel-Borowski K (1992) The interphase microtubule cytoskeleton of five different phenotypes of microvessel endothelial cell cultures derived from bovine corpus luteum. Tissue Cell 24:347–354
Wright JW, Pejovic T, Lawson M, Jurevic L, Hobbs T, Stouffer RL (2010) Ovulation in the absence of the ovarian surface epithelium in the primate. Biol Reprod 82:599–605
Wu R, van der Hoek KH, Ryan NK, Norman RJ, Robker RL (2004) Macrophage contributions to ovarian function. Hum Reprod Update 10:119–133
Wu R, Fujii S, Ryan NK, van der Hoek KH, Jasper MJ, Sini I, Robertson SA, Robker RL, Norman RJ (2007) Ovarian leukocyte distribution and cytokine/chemokine mRNA expression in follicular fluid cells in women with polycystic ovary syndrome. Hum Reprod 22:527–535
Yacobi K, Tsafriri A, Gross A (2007) Luteinizing hormone-induced caspase activation in rat preovulatory follicles is coupled to mitochondrial steroidogenesis. Endocrinology 148:1717–1726
Zeleznik AJ (1993) Dynamics of primate follicular growth: a physiologic perspective. In: Adashi EY, Leung PCK (eds) The ovary. Raven, New York, pp 41–55
Zhan R, Leng X, Liu X, Wang X, Gong J, Yan L, Wang L, Wang Y, Wang X, Qian LJ (2009) Heat shock protein 70 is secreted from endothelial cells by a non-classical pathway involving exosomes. Biochem Biophys Res Commun 387:229–233
Zhou M, McFarland-Mancini MM, Funk HM, Husseinzadeh N, Mounajjed T, Drew AF (2009) Toll-like receptor expression in normal ovary and ovarian tumors. Cancer Immunol Immunother 58:1375–1385
Zovein AC, Hofmann JJ, Lynch M, French WJ, Turlo KA, Yang Y, Becker MS, Zanetta L, Dejana E, Gasson JC, Tallquist MD, Iruela-Arispe ML (2008) Fate tracing reveals the endothelial origin of hematopoietic stem cells. Cell Stem Cell 3:625–636

Index

U

V

Printing: Ten Brink, Meppel, The Netherlands
Binding: Stürtz, Würzburg, Germany